T0137223

Studies in Computational Intelligence

Volume 764

Series editor

Janusz Kacprzyk, Polish Academy of Sciences, Warsaw, Poland
e-mail: kacprzyk@ibspan.waw.pl

The series "Studies in Computational Intelligence" (SCI) publishes new developments and advances in the various areas of computational intelligence—quickly and with a high quality. The intent is to cover the theory, applications, and design methods of computational intelligence, as embedded in the fields of engineering, computer science, physics and life sciences, as well as the methodologies behind them. The series contains monographs, lecture notes and edited volumes in computational intelligence spanning the areas of neural networks, connectionist systems, genetic algorithms, evolutionary computation, artificial intelligence, cellular automata, self-organizing systems, soft computing, fuzzy systems, and hybrid intelligent systems. Of particular value to both the contributors and the readership are the short publication timeframe and the world-wide distribution, which enable both wide and rapid dissemination of research output.

More information about this series at http://www.springer.com/series/7092

Rafael Valencia-García
Mario Andrés Paredes-Valverde
María del Pilar Salas-Zárate
Giner Alor-Hernández
Editors

Exploring Intelligent Decision Support Systems

Current State and New Trends

 Springer

Editors
Rafael Valencia-García
Departamento de Informática y Sistemas,
    Facultad de Informática
Universidad de Murcia
Murcia
Spain

Mario Andrés Paredes-Valverde
Departamento de Informática y Sistemas,
    Facultad de Informática
Universidad de Murcia
Murcia
Spain

María del Pilar Salas-Zárate
Departamento de Informática y Sistemas,
    Facultad de Informática
Universidad de Murcia
Murcia
Spain

Giner Alor-Hernández
Division of Research and Postgraduate
    Studies
Instituto Tecnológico de Orizaba
Orizaba, Veracruz
Mexico

ISSN 1860-949X   ISSN 1860-9503   (electronic)
Studies in Computational Intelligence
ISBN 978-3-319-89266-5   ISBN 978-3-319-74002-7   (eBook)
https://doi.org/10.1007/978-3-319-74002-7

This Springer imprint is published by Springer Nature
The registered company is Springer International Publishing AG
The registered company address is: Gewerbestrasse 11, 6330 Cham, Switzerland

Preface

Decision Support Systems (DSSs) are gaining an increased popularity in various domains, including business, engineering, the military, and medicine. They are especially valuable in situations in which the amount of available information is prohibitive for the intuition of an unaided human decision-maker and in which precision and optimality are of importance. DSS can aid human cognitive deficiencies by integrating various sources of information, providing intelligent access to relevant knowledge, and aiding the process of structuring decisions. They can also support choice among well-defined alternatives and build on formal approaches, such as the methods of engineering economics, operations research, statistics, and decision theory. Furthermore, they employ artificial intelligence methods to address heuristically problems that are intractable by formal techniques. Proper application of decision-making tools increases productivity, efficiency, and effectiveness and gives many businesses a comparative advantage over their competitors, allowing them to make optimal choices for technological processes and their parameters, planning business operations, logistics, or investments. These systems allow individuals and organizations to deal with unstructured or semi-structured decision problems that demand extensive experience and expert knowledge.

The field of DSS is expanding to use new technologies such as Social Media, Semantic Web, Linked Data, Big Data, Machine Learning, Geographic Information System, etc. These technologies are converging to provide integrated support for individuals and organizations to make more rational decisions in many fields such as Agriculture, Agrotechnology, Clinical, eCommerce, eGovernment, eHealth, eLearning, Information Retrieval, Supply Chain, Water Resource Management, etc.

The aim of this book is to explore the latest results of research, development, and applications of DSS in different fields as well as to disseminate innovative and high-quality research regarding the foundations, methods, methodologies, models, tools, and techniques for designing, developing, implementing, and evaluating advanced DSS in different fields.

- Create a collection of theoretical, real-world, and original research works in the field of DSS.
- Go beyond the state of the art in the field of DSS.
- Publish successful applications and use cases of new approaches, applications, methods, and techniques for developing advanced DSS and their application in different fields.
- Provide an appropriate dissemination venue for both academia and industrial communities.

This book contains two kinds of contribution: regular research papers and case studies. These works have been edited according to the norms and guidelines of Springer Verlag Editorial. Several calls for chapters were distributed among the main mailing lists of the field for researchers to submit their works to this issue. In the first deadline, we received a total of 29 expressions of interest in the form of abstracts. Due to the large amount of submissions, abstracts were subject to a screening process to ensure their clarity, authenticity, and relevancy to this book. Proposals came from several countries such as Brazil, Colombia, Ecuador, Germany, India, Mexico, Norway, Spain, Taiwan, United Kingdom of Great Britain, Northern Ireland, and United States of America.

After the screening process, 16 proposals were invited to submit full versions. At least two reviewers were assigned to every work to proceed with the peer review process. Eleven chapters were finally accepted for their publication after corrections requested by reviewers and editors were addressed.

The book content is structured in two parts: (1) DSS for Industry and (2) Case Studies: Clinical, Emergency Management, and Pollution Control.

DSS for Industry: this part contains six chapters.

Chapter entitled "FINALGRANT: A Financial Linked Data Graph Analysis and Recommendation Tool" presents FINALGRANT, an alternative solution for analyzing and visualizing digital financial data. This solution retrieves XBRL-based financial data and transforms them into RDF triples through a linked data-based approach. Furthermore, FINALGRANT addresses the lack of a semantic property to interlink the data and make it available, as well as the challenge of accessing information through Internet protocols such as HTTP.

Chapter entitled "Constructing and Interrogating Actor Histories" presents a novelty approach for creating and analyzing actor histories. This approach allows defining precisely how the execution histories are produced and gives a complete specification of the query language that can be used to analyze them. The proposed approach was evaluated in a shop where customers browse for items, seek help from assistants, and queue to buy chosen items. The shop is interested in how to organize its assistants, sales, and floor-walking strategies to minimize unhappy customers.

Chapter entitled "Challenges in the Design of Decision Support Systems for Port and Maritime Supply Chains" presents a systematic literature review on the design and development of DSS for the port and maritime industry. The authors review analytical and technological methods, and distinguish the gaps and tendencies of future development in such domain. Finally, they conclude that DSS in the context of maritime transport will take advantage of collaborative systems and data analytics to improve decision-making processes.

Chapter entitled "Analyzing the Impact of a Decision Support System on a Medium Sized Apparel Company" presents the system analysis of a medium-sized apparel company that was performed to establish the requirements of a DSS as well as the effects of using it in the company. The DSS was implemented for 1 year, achieving results such as the improvement of user performance and response time to consumers, as well as the increasing of financial turnover rate, customer service, and decision quality.

Chapter entitled "A Multicriteria Decision Support System Framework for Computer Selection" proposes a theoretical framework that allows families to evaluate computers from a multi-attribute perspective by using the Analytic Hierarchy Process (AHP) and the Technique for Order of Preference by Similarity to Ideal Solution (TOPSIS). This framework was evaluated through a case study, and the findings demonstrate that it is friendly to users since they can perform the evaluation process on their own.

Chapter entitled "An Agent-Based Memetic Algorithm for Solving Three-Level Freight Distribution Problems", presents an agent-based memetic algorithm that combines global and local strategies for solving a three-level freight distribution network. This solution uses coordination and collaboration strategies between several agents to improve the performance of the freight transport process. The proposed solution obtained better results than Solomon insertion heuristic, in terms of less total travel distance and variability.

Case Studies: Clinical, Emergency Management, and Pollution Control: this part contains five chapters.

Chapter entitled "DiabSoft: A System for Diabetes Prevention, Monitoring, and Treatment" presents DiabSoft, a medical DSS that monitors patient health parameters mainly to obtain data on their vital signs, daily habits, and symptoms, and then generate recommendations based on collaborative filtering and shared knowledge. The system proposed relies on data interchange and integration to generate medical recommendations for patients and healthcare professionals.

Chapter entitled "Health Monitor: An Intelligent Platform for the Monitorization of Patients of Chronic Diseases" proposes Health Monitor, a platform for the self-management of chronic diseases, more specifically for diabetes mellitus and hypertension is presented. This platform provides health recommendations based on the data collected by the motorization process. Health Monitor was evaluated by real patients with diabetes mellitus and hypertension. The evaluation results show that Health Monitor provides recommendations that can enhance patients' quality of life.

Chapter entitled "Reliable and Smart Decision Support System for Emergency Management Based on Crowdsourcing Information" describes the RESCUER system, a smart and interoperable DSS for emergency and crisis management based on mobile crowdsourcing information. RESCUER was evaluated with end users obtaining quite positive results.

Chapter entitled "Intelligent Decision Support for Unconventional Emergencies" presents a set of best practices for developing a DSS for emergency management. The chapter focuses upon effective and efficient storage, analysis and processing of emergency information, safety plans and resources, and applying a rule-based approach to generate the recommendations in case of emergency along with proper justification. Furthermore, an ontology representation scheme has been used to represent human knowledge and reason with it.

Chapter entitled "Information Technology in City Logistics: A Decision Support System for Off-Hour Delivery Programs" proposes an information technology strategy based on a DSS for gathering and sharing data about night deliveries. This strategy was implemented in Bogotá (Colombia) showing that designing a robust outreach OHD (Off-Hour Delivery) program requires the identification and synchronization of all players involved in urban deliveries and establishing the institutional framework for freight issues in city agencies.

Once a brief summary of chapters has been provided, we would also like to express our gratitude to the reviewers who kindly accepted to contribute in the evaluation of chapters at all stages of the editing process.

Murcia, Spain   Rafael Valencia-García
Murcia, Spain   Mario Andrés Paredes-Valverde
Murcia, Spain   María del Pilar Salas-Zárate
Orizaba, Mexico   Giner Alor-Hernández

Acknowledgements

Guest editors will always be grateful for the talented technical reviewers who helped review and improve this book. The knowledge and enthusiasm they brought to the project were simply amazing.

Thus, we would like to thank all our colleagues and friends from the Instituto Tecnológico de Orizaba and Universidad de Murcia for all their support.

We equally and especially wish to thank Springer Verlag and associate editors of Studies in Computational Intelligent book series, for grating us the opportunity to edit this book and providing valuable comments to improve the selection of research works.

Guest editors are grateful to the National Technological of Mexico for supporting this work. This book was also sponsored by the National Council of Science and Technology (CONACYT) as part of the project named Thematic Network in Industrial Process Optimization, as well as by the Public Education Secretary (SEP) through PRODEP. Finally, this book has been partially supported by the Spanish National Research Agency (AEI) and the European Regional Development Fund (FEDER/ERDF) through project KBS4FIA (TIN2016-76323-R).

Contents

**Part II   Case Studies: Clinical, Emergency Management and
          Pollution Control**

Contributors

Wilson Adarme Jaimes Universidad Nacional de Colombia, Bogotá D.C., Colombia

Giner Alor-Hernández Division of Research and Postgraduate Studies, Instituto Tecnológico de Orizaba, Orizaba, Veracruz, Mexico

Oscar Apolinario Cdla. Universitaria Salvador Allende, Universidad de Guayaquil, Guayaquil, Ecuador

Martín Darío Arango-Serna Universidad Nacional de Colombia—Sede Medellin, Medellín, Antioquia, Colombia

Souvik Barat Tata Consultancy Services Research, Pune, India

Balbir Barn Middlesex University, London, UK

Norberto Castillo-García Faculty of Engineering, Universidad Autonoma de Tamaulipas, Tampico, Mexico

Juan Pablo Castrellón-Torres Universidad Nacional de Colombia, Bogotá D.C., Colombia

Tony Clark Sheffield Hallam University, Sheffield, UK

Nancy Aracely Cruz-Ramos Division of Research and Postgraduate Studies, Instituto Tecnológico de Orizaba, Orizaba, Veracruz, Mexico

Roberto Díaz-Reza Department of Electric and Computational Sciences, Universidad Autónoma de Ciudad Juárez, Ciudad Juárez, Chihuahua, Mexico

P. Erhan Eren Information Systems, Informatics Institute, Middle East Technical University, Ankara, Turkey

Tobias Franke German Research Center for Artificial Intelligence (DFKI), Kaiserslautern, Germany

Jorge Luis García-Alcaraz Department of Industrial Engineering and Manufacturing, Universidad Autónoma de Ciudad Juárez, Ciudad Juárez, Chihuahua, Mexico

María D. Gracia Faculty of Engineering, Universidad Autonoma de Tamaulipas, Tampico, Mexico

Cristian Giovanny Gómez-Marín Universidad Nacional de Colombia—Sede Medellin, Medellín, Antioquia, Colombia

Ebru Gökalp Information Systems, Informatics Institute, Middle East Technical University, Ankara, Turkey

Mert Onuralp Gökalp Information Systems, Informatics Institute, Middle East Technical University, Ankara, Turkey

Sarika Jain Department of Computer Applications, National Institute of Technology, Kurukshetra, Haryana, India

Vinay Kulkarni Tata Consultancy Services Research, Pune, India

Jorge Kurano Polytechnic University of Madrid, Madrid, Spain

Katty Lagos-Ortiz Cdla. Universitaria Salvador Allende, Universidad de Guayaquil, Guayaquil, Ecuador

Harry Luna-Aveiga Cdla. Universitaria Salvador Allende, Universidad de Guayaquil, Guayaquil, Ecuador

Néstor Eliécer Manosalva Barrera Universidad Nacional de Colombia, Bogotá D.C., Colombia

Julio Mar-Ortiz Faculty of Engineering, Universidad Autonoma de Tamaulipas, Tampico, Mexico

Valeria Martínez-Loya Department of Industrial Engineering and Manufacturing, Universidad Autónoma de Ciudad Juárez, Ciudad Juárez, Chihuahua, Mexico

José Medina-Moreira Cdla. Universitaria Salvador Allende, Universidad de Guayaquil, Guayaquil, Ecuador

Jose Manuel Menendez Polytechnic University of Madrid, Madrid, Spain

Claudia Nass Fraunhofer Institute for Experimental Software Engineering, Kaiserslautern, Germany

Renato Novais Fraunhofer Project Center for Software and Systems Engineering, Federal University of Bahia, Instituto Federal de Educacão, Ciência e Tecnologia da Bahia, Salvador, Brazil

Mario Andrés Paredes-Valverde Departamento de Informática y Sistemas, Facultad de Informática, Universidad de Murcia, Murcia, Spain; Division of Research and Postgraduate Studies, Instituto Tecnológico de Orizaba, Orizaba, Veracruz, Mexico

Andreas Poxrucker German Research Center for Artificial Intelligence (DFKI), Kaiserslautern, Germany

Jose Rodrigues Jr. University of Sao Paulo, Sao Carlos, Brazil

Lisbeth Rodríguez-Mazahua Division of Research and Postgraduate Studies, Instituto Tecnológico de Orizaba, Orizaba, Veracruz, Mexico

María del Pilar Salas-Zárate Division of Research and Postgraduate Studies, Instituto Tecnológico de Orizaba, Orizaba, Veracruz, Mexico

Conrado Augusto Serna-Urán Universidad de San Buenaventura, Medellín, Antioquia, Colombia

Paulo Simões Jr. Fraunhofer Project Center for Software and Systems Engineering, Federal University of Bahia, Salvador, Brazil

Liliana Avelar Sosa Department of Industrial Engineering and Manufacturing, Universidad Autónoma de Ciudad Juárez, Ciudad Juárez, Chihuahua, Mexico

José Luis Sánchez-Cervantes CONACYT-Instituto Tecnológico de Orizaba, Orizaba, Veracruz, Mexico

José Sebastián Talero Chaparro Universidad Nacional de Colombia, Bogotá D.C., Colombia

Jairo Humberto Torres Acosta Universidad Nacional de Colombia, Bogotá D.C., Colombia

Agma Traina University of Sao Paulo, Sao Carlos, Brazil

Ismael Canales Valdiviezo Department of Electric and Computational Sciences, Universidad Autónoma de Ciudad Juárez, Ciudad Juárez, Chihuahua, Mexico

Rafael Valencia-García Facultad de Informática, Universidad de Murcia, Murcia, Spain

Karina Villela Fraunhofer Institute for Experimental Software Engineering, Kaiserslautern, Germany

Julián Andrés Zapata-Cortés Institución Universitaria CEIPA, Sabaneta, Antioquia, Colombia

List of Figures

List of Tables

Key Points

- **How the book approaches its subject matter:**

  DSS aids human cognitive deficiencies by integrating various sources of information, providing access to relevant knowledge, and aiding the process of structuring decisions. The field of DSS is expanding to use new technologies such as Social Media, Semantic Web, Linked Data, Big Data, Machine Learning, Geographic Information System, etc.

- **What is new about this approach:**

  This book considers innovative and high-quality research for developing advanced DSS-based methods and techniques and their application in the following industrial sectors: Apparel Industry, Port and Maritime Industry, eHealth, Clinical, Emergency Management, Pollution Control, and others.

- **The book's general scope:**

  The goal of this book is to disseminate current trends among innovative and high-quality research regarding the implementation of conceptual frameworks, strategies, techniques, methodologies, informatics platforms, and models for developing advanced DSS and their application in different fields.

- **Intended readership:**

  The subject area is a combination of different fields of Artificial Intelligence by applying knowledge representation, actor histories, memetic algorithm, Semantic Web, and TICs used in different domains in which a broad scope of audience is interested in

  - Stakeholders.
  - Analyst.
  - Consultant.

- Professors in academia.
- Students.
- Corporate heads of firms.
- Senior general managers.
- Managing directors.
- Board directors.
- Academics and researchers in the field both in universities and business schools.
- Information technology directors and managers.
- Quality managers and directors.
- Libraries and information centers serving the needs of the above.

- **Any key uses.**

The book contents are also useful for Ph.D., master, and undergraduate students of IT-related degrees such as Computer Science, Information Systems, and Electronic Engineering.

Part I
Decision Support Systems for Industry

FINALGRANT: A Financial Linked Data Graph Analysis and Recommendation Tool

José Luis Sánchez-Cervantes, Giner Alor-Hernández,
María del Pilar Salas-Zárate, Jorge Luis García-Alcaraz
and Lisbeth Rodríguez-Mazahua

Abstract Current digital financial information is generated by data located in a distributed but linked environment. In this sense, semantic technologies and linked data allow files and data alike to be first-class web resources and promote information distribution and knowledge sharing within a global, open-standard space known as the Data Web. This chapter proposes FINALGRANT as an alternative solution to analyzing and visualizing digital financial data. The tool retrieves XBRL-based financial data and transforms them into RDF triples through a process inspired in the linked data principles. FINALGRANT identifies and addresses some limitations of financial statements, such as the lack of a semantic property to interlink the data and make it navigable and the challenge of accessing information through Internet protocols, such as HTTP, to navigate among data and interconnect them with external data sources. Similarly, FINALGRANT can search for financial ratios and processes fundamental or classical analysis calculations to support fund investment decisions.

Keywords Financial data · Linked data · XBRL

J. L. Sánchez-Cervantes (✉)
CONACYT-Instituto Tecnológico de Orizaba, Av. Oriente 9
no. 852 Col. E. Zapata, CP 94320 Orizaba, Veracruz, Mexico
e-mail: jlsanchez@conacyt.mx

G. Alor-Hernández · M. d. P. Salas-Zárate · L. Rodríguez-Mazahua
Division of Research and Postgraduate Studies, Instituto Tecnológico de Orizaba,
Av. Oriente 9 no. 852 Col. E. Zapata, CP 94320 Orizaba, Veracruz, Mexico

J. L. García-Alcaraz
Department of Industrial Engineering and Manufacturing, Universidad Autónoma
de Ciudad Juárez, Av. del Charro 450 Norte, Ciudad Juárez 32310, Chihuahua, Mexico

© Springer International Publishing AG 2018
R. Valencia-García et al. (eds.), *Exploring Intelligent Decision Support Systems*,
Studies in Computational Intelligence 764,
https://doi.org/10.1007/978-3-319-74002-7_1

1 Introduction

The current financial data ecosystem benefits from the increasing amount of information that corporations constantly disclose on the Web. Many financial laws and regulations demand publicly owned companies to disclose certain business and financial data on a regular basis to protect investors and customers from possible frauds by businesses. On example of these laws is the *Sarbanes-Oxley*[1] act, a federal law enacted for the United States' Securities and Exchange Commission[2] to encourage corporate responsibility and transparency and avoid corporate and accounting fraud.

To comply with financial laws and regulations and obtain all the possible benefits, corporations need effective financial data analysis solutions. In this sense, the importance of financial data exchange relies on data reutilization, which can eventually encourage the development of data analysis tools that efficiently issue, receive, consolidate, and analyze digital financial information.

The *eXtensible Business Reporting Language* (XBRL)[3] initiative [1] promotes a standardized approach to digital financial reporting for companies that rely on XML-based technologies [2] to disclose financial and business data. Some of the tools used by XBRL include balance sheets, cash flow statements, and income statements. In accounting, a balance sheet, or a general balance sheet, is described as "*the instantaneous report of the financial situation of a company*" [3].

The XBRL standard plays a key role in digital financial reporting and paves the way to new forms of data exchange among computer applications [1]. However, the language itself cannot solve data integration problems, because it lacks the semantics to keep data interrelated [4]. That is, the data are not navigable among them and cannot be interconnected with external data sources. As a result, XBRL offers limited opportunities for companies to perform tasks such as creating a financial knowledge base for rule-based inferences and applying financial data science.

To address the problem of data integration, we propose FINALGRANT (FInancial Linked Data GRaph ANAlysis Tool), a tool for calculating, analyzing, and generating graphs on financial information, namely financial ratios. FINALGRANT relies on SPARQL-based queries performed on a financial knowledge base previously generated after transforming financial statements in XBRL into linked open data (semantic annotation). In fact, linked data allows developers to exploit the multiple connections existing among digital financial data.

The remainder of this paper is a follows: Sect. 2 discusses the basic aspects of financial analysis, including financial ratios, financial statements, and XBRL as the standard for digital financial reporting. Then, Sect. 3 describes FINALGRANT's process of transforming XBRL data into linked data. A case study is presented in

---

[1]Sarbanes-Oxley: http://www.sec.gov/about/laws.shtml#sox2002.

[2]U.S. SEC: http://www.sec.gov/.

[3]XBRL: http://www.xbrl.org/.

Sect. 4 of this paper to depict the functionality of FINALGRANT. Finally, Sect. 5 presents the conclusions and suggestions for future research.

2 Fundamental Analysis

Fundamental analysis determines the value of corporate securities through a careful examination of key-value drivers or factors [5], such as earnings, investments, risk, growth, and competitive position. In this sense, financial ratios are a key in fundamental analysis because they help decision makers to accurately calculate corporate performance and show a business' financial health [6]. Therefore, it can be said that the goals of fundamental analysis are to determine a company's financial health by analyzing its current and past financial ratios and to make financial preventions based on the obtained results.

2.1 Financial Ratios

Financial ratios are key indicators of a company's financial health [7] and can determine the types of financial changes that the corporation must undergo at some time. There are four main groups of financial ratios [8]: liquidity ratios, activity and asset management ratios, financial leverage or debt ratios, and profitability ratios. **Liquidity ratios** examine a company's ability to pay its short-term obligations. **Activity and asset management ratios** compare assets and net sales with total assets, fixed tangible assets, current assets, or their elements. **Financial leverage or debt ratios** relate a company's resources with its obligations. **Profitability ratios** measure a company's ability to generate economic and financial wealth.

Financial ratios can also be divided in more specific groups [8, 9]: turnover ratios, liquidity ratios, and solvency ratios. Turnover ratios, also known as efficiency or activity ratios, show a company's ability to manage its expenses and generate revenue and cash from its resources. That said, there are six types of turnover ratios: assets turnover, non-current assets turnover, current assets turnover, customer turnover, stock turnover, and capital turnover. On the other hand, liquidity ratios indicate a company's ability to turn its assets into cash and measure its ability to pay its short-term obligations. There are seven liquidity ratios: working capital, net working capital over total assets, net working capital on short-term debt, acid test, day time interval measurement, current ratio, and cash ratio. Finally, solvency ratios diagnose whether a company is able to timely and orderly pay off its debts. There are two solvency ratios: the debt ratio and the debt quality ratio.

All the aforementioned financial indicators can be calculated with FINALGRANT. Table 1 summarizes the financial ratios discussed in the previous subsection. For every ratio, the table presents its definition and formula and specifies its usefulness in corporate contexts. Also, the last part of the table includes

Table 1 Turnover ratios [8, 9]

Ratio	Formula	Definition	Usefulness
Turnover ratios			
Total assets turnover	NS/TA	Measures how many times the net sales cover the total assets during a particular period. In other words, the total assets turnover ratio indicates how efficiently a company uses its assets to generate sales	Allows companies to know how many sales are generated from the total assets. Based on this ratio, assets may increase or decrease
Non-current assets turnover	NS/NCA	Obtained by dividing the net sales revenue between the non-current assets, which include fixed tangible and intangible assets, investments, financial credits, and real estate investments	Reflects how many times the non-current assets of a company have been used to generate revenue. It is an efficiency ratio that shows how efficiently a business is in managing its non-current assets. Companies usually expect this ratio to be as high as possible
Current assets turnover	NS/CA	It is calculated by dividing the net sales between the average current assets. Current assets include debtors, investments, financial credits expiring in less than one year, and treasury, among others	Indicates how many sales a company generates thanks to its current assets. Based on this ratio, current assets might increase or decrease
Account receivable turnover	NS/AR	It measures the turnover of account receivables, and it is used to evaluate the payment terms that a company offers its customers	Measures how many times a company collects its average accounts receivable. Companies usually expect this ratio to be as high as possible, since it means they can quickly convert their receivables into cash
Stock turnover	NS/I	Indicates the number of times that a company converts its stock into sales during a certain time period	One of the most used metrics in supply chain. Companies usually expect this ratio to be as high as possible. A low stock turnover ratio implies the stock is taking longer to turn over
Capital turnover	NS/WC = NS/ Operational Existing Revolving Fund	Compares the depletion of working capital used to fund operations and purchase inventory	Should be as high as possible. Companies usually compare this metric with the sector's representative value. Also known as equity turnover ratio, the capital turnover ratio allows companies to know the proportion of their sales to their shareholders' equity

(continued)

Table 1 (continued)

Ratio	Formula	Definition	Usefulness
Liquidity ratios			
Working capital	CA − CL	Indicates whether a company has enough assets to cover its short-term debt. The working capital ratio is the difference between a company's current assets (in less than a fiscal year) and current liabilities	A positive WC ratio encourages investment. On the other hand, a negative working capital ratio indicates the company needs to rely on borrowing or stock issuance to finance its working capital and pay its current debts. The financial aid the company would receive as consequence must make the ratio go from negative to positive and must help the business pay its current obligations
Net working capital over total assets	(CA − CL)/TA	Refers to the net working capital of a company as a percentage of its total assets. It measures the relationship between a company's net working capital and total assets	The optimal working capital over total assets ratio is any value above 0, since a low ratio indicates low liquidity levels (not the appropriate number of current assets). The interpretation of this ratio depends on the sector wherein the company operates
Net working capital on short-term debt	(CA − CL)/CL	It is calculated by dividing a company's net working capital (of no more than a fiscal year) by its current short-term debts or liabilities	The optimal ratio must be close to 0.5, otherwise it indicates that the company struggles to cover its short-term debts, even though it coverts all its assess into cash
Acid test	(CA − I)/CL	Reveals whether a company has enough short-term assets to cover its immediate liabilities. This ratio is calculated by dividing a company's current assets—without considering inventory—by its immediate liabilities	Reveals a company's ability to cover its immediate liabilities without considering inventory, supplies, and prepared expenses. A less than 1 acid-test ratio indicates that the company lacks enough liquid assets to pay its debt. On the other hand, a too high ratio could indicate an excess of liquidity (i.e. cash has accumulated and is isle.) The optimal acid test ratio ranges from 0.5 to 1
Day time interval measurement	(CA − CM) * 365	This ratio is the quotient of a company's current assets divided by its current materials, multiplied by 365 days	Measures for how many more days the company can operate within a year in case its daily activities are abruptly paralyzed

(continued)

Table 1 (continued)

Ratio	Formula	Definition	Usefulness
Current ratio	CA/CL	Measures a company's ability to pay its short-term and long-term obligations	It is the main measure of liquidity. The higher the ratio, the higher the company's solvency and ability to pay its obligations
Cash ratio	CE/CL	Shows the relationship between a company's cash and cash equivalents to its total current liabilities. Cash equivalents are assets that can be quickly converted into cash (i.e. in one or two days). Therefore, they do not include accounts receivable. On other hand, current liabilities are to be settled within 12 months or the company's cycle	The optimal cash ratio is 0.3, meaning that for every dollar of current liabilities the company has 0.3 dollars in cash within two or three days
Solvency ratios			
Debt to equity ratio	TL/SE	Measures a company's financial leverage. It is calculated by dividing the company's total liabilities by its stakeholders' equity. There is not an ideal debt/equity ratio, as it depends on each company	Can be interpreted as follows: (A) A ratio of 1 indicates that stockholders and creditors make equal contributions to the business. (B) A less than 1 ratio indicates that the proportion of assets from stockholders is greater than the proportion from creditors. (C) A higher than 1 ratio implies that creditors provide more assets than stockholders. Note that a less than 1 ratio is not always detrimental, as it may be an indicator of greater protection to the company's money
Debt quality ratio	STL/TL	Calculated by diving a company's short-term liabilities by its total liabilities	A company's debts are of better or worse quality depending on their duration. The more short-time debts, the worse, as this means that the company would have to divest of economic resources to pay the debt very soon. That said, the smaller the quality ratio, the better the debt's quality

(continued)

Table 1 (continued)

Ratio	Formula	Definition	Usefulness
Debt ratio	((TL/TA) * 100)	Shows the degree of participation from stakeholders in a company's assets. Can be defined as the ratio of total liabilities to total assets, expressed as a decimal percentage	A debt ratio of 60% is manageable, as it means that for every 100.00 USD of assets, a company has 60.00 USD worth of liabilities

CA Current assets; *CL* current liabilities; *CM* cost of materials; *CE* cash and cash receivables; *WC* working capital; *NS* net sales; *STL* short-term liabilities; *TL* total liabilities; *AR* account receivable; *SE* stakeholders' equity; *TA* total assets; *NCA* non-current assets the abbreviations and acronyms used in the formulas. Note that the ratio names and formulas follow the financial taxonomy set by the US-GAAP (Generally Accepted Accounting Principles) norm.

2.2   Financial Statements

A financial statement is a formal record that details the financial activities of a business during a given period. Usually, companies submit quarterly, semi-annual, and annual financial statements. In the United States, the Securities and Exchange Commission (SEC) stores and discloses the financial statements of public American companies through the Electronic Data Gathering Analysis and Retrieval System (EDGAR). Companies must submit their financial statements by using EDGAR XBRL filings that comply with certain regulations and forms, such as the 10-K annual report form, the N-SAR semi-annual report form, and the 10-Q quarterly report form [10]. In this research, we will deal merely with quarterly reports, since they are said to offer a continuous view a company's financial position during the course of a year.

There are two main types of data to analyze and calculate financial information: real-time data and public disclosure data. In turn, public disclosure data can be divided into three types of financial statements [11]:

1. **Balance Sheet**: Shows the financial situation of a company at a specific point in time. Balance sheets summarize a company's assets, liabilities, and shareholders' equity. Companies listed in the New York Stock Exchange must file quarterly balance sheets.
2. **Income Statement**: Shows the results of a company's operations for a given time period. Specifically, income statements summarize a company's revenues, gains, expenses, and losses for a given time.
3. **Cash Flow Statement**: Details the sources of a company's cash on a given time and states how such cash was spent that time period.

2.3   Financial Statements in XBRL

XBRL is composed of two parts: a taxonomy and the XBRL-based statements. The taxonomy defines the reporting concepts and includes additional information regarding such concepts (e.g. labels in multiple languages, interrelationships between concepts, validation formulas, references to legal norms). XBRL tax-onomies are like the rules that verify the content of the form. On the other hand, the XBRL-based statements include the taxonomies; and the financial information is entered by using the concepts previously defined [12].

An XBRL taxonomy is formed by the definition of an XML schema through an SML Schema Definition (XSD) [13, 14] and the XBRL linkbases, which are the

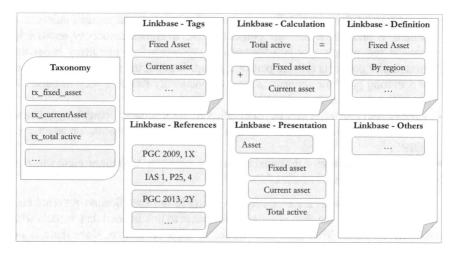

Fig. 1 Structure of the taxonomies for financial statements disclosed in XBRL

physical XML files that contain information about the elements defined in the XBRL Schema. Also, an XBRL taxonomy can be part of a set of related taxonomies, known as a Discoverable Taxonomy Set (DTS) [15]. The XML Schema in a taxonomy defines the reporting concepts, which are assigned a name and type as two definitions of XML Schema elements.

Figure 1 depicts the structure of the taxonomies for financial statements disclosed in XBRL.

According to the communication theory, for a message to be properly transmitted, both the sender and the receiver must share a code, which is the form in which the message is sent. In XBRL, taxonomies can be considered as the code. As previously mentioned, XBRL taxonomies are composed of XML Schemas [12] that define the set of elements present in the financial statements and the structure of such elements. This set of elements is also known as the dictionary of reporting-area specific terms, and these terms, as depicted in Fig. 1, have a specific function [12, 15]:

- **Label linkbase**: Contains the labels associated to the dictionary elements in multiple languages. Humans interpret the data that corresponds to the concept appearing in the same row. For instance: *Fixed assets* $150,000.
- **Reference linkbase**: These are references to authoritative statements published in business, financial, and accounting literature. Such references play a key role when it comes to clarifying the use of reporting concepts. Also, this linkbase is useful when locating the concepts to be used to prepare reports in XBRL that use a standard taxonomy, such as the International Financial Reporting Standards (IFRS) or US-GAAP.

- **Presentation linkbase**: Defines the rules for building the representation of a statement and has a twofold purpose. First, it allows taxonomy authors to arrange sets of concepts into hierarchical representations to convey better the meaning of concepts that are part of a group. Second, the presentation linkbase is a starting point for applications that automatically format reports to build the templates that will show the data.
- **Calculation linkbase**: Sets the calculation rules (arithmetic operations) between the taxonomy elements that validate the financial statements or reports in XBRL.
- **Definition linkbase**: Additional rules that correlate taxonomy elements that explain or document relationships.

The XBRL linkbases are extensible. Nothing in the specification prevents taxonomy developers to develop proprietary linkbases to link internal data models with taxonomy elements. That said, such linkbases must be private, since there is no approved specification in the consortium for all XBRL processors to understand them [15].

3 The XBRL–Linked Data Transformation Process with FINALGRANT

FINALGRANT's financial data extraction and interconnection process involves transforming the data stored in the XBRL-based files in RDF triples to generate a searchable, computable financial linked dataset. This transformation process consists of six stages depicted in Fig. 2 and thoroughly discussed below:

(1) **Data acquisition**: A crawler is used to track financial information sources that have quarterly financial reports (10-Q form) disclosed under the XBRL standard [10]. Then, the financial statements are downloaded and stored for further transformation in RDF triples.
(2) **Semantic model**: The semantic model defines the semantic relationships between the concepts and conceptual classes that capture all the financial data aspects expressed in the XBRL-based files. Additionally, the model allows performing analytical procedures with crossed financial statements and processing SPARQL-based queries. Another important aspect of the semantic model is its capability to reutilize semantic vocabulary. For instance, the Time Ontology is used to represent financial data reporting periods, whereas the Payments Ontology is used for representing organizational spending information in linked data.
(3) **Data extraction**: This stage involves retrieving the data from of the XBRL-based financial statements obtained in the data acquisition stage. The financial statements include balance sheets, income statements, cash flow statements, and additional relevant information. Also, the financial data are

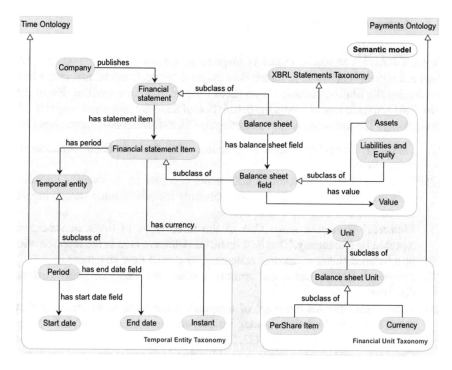

Fig. 2 Linked-data inspired semantic model for financial data

extracted along with the XBRL taxonomies and their linkbases (i.e. labels, references, definitions, and calculations). Finally, for every financial statement, complementary metadata such as dates, company information, and financial statement reporting periods, are equally retrieved.

(4) **Semantic dataset generation**: This stage is interrelated with the data extraction stage. At this stage, data are serialized semantically according to the definition of the semantic model. The data serialization is performed using Apache Jena's framework [16], and the financial data are stored in the form of RDF triples within the Virtuous Open-Source semantic repository [17].

(5) **Data integration**: Once the semantic dataset is loaded with the RDF triples, its semantic structure is significantly reduced to the barriers of linking the data with other datasets. In the context of linked data, this process is known as data interlinking [18], in which local dataset concepts are interlinked with the concepts of external data sources.

(6) **SPARQL-based queries**: Once the data have been extracted and interlinked, they can be analyzed. SPARQL-based queries are used to perform meta-analyses of the calculated ratios and to provide FINALGRANT with a proper data calculation framework.

3.1 Semantic Model for Financial Data

FINALGRANT's semantic model is inspired in the linked data principles [19]. Figure 2 represents the mixed linked-data-inspired semantic model that arises from combining the Entity-Attribute-Value (EAV) model and the canonical design pattern. This semantic model favors the calculation of financial ratios and supports data navigability. The following sections thoroughly discuss each model component [20]:

(1) *Company*: The public company or business that discloses financial statements on the Web.
(2) *Financial Statement*: Official documents reported by companies regarding their financial situation. Financial statements include balance sheets, income statements, and cash flow statements.
(3) *Financial Statement Item*: One of the three types of financial statements reported by a company. That is, a financial statement item is the representation of a balance sheet, an income statement, or a cash flow statement.
(4) *Temporal Entity*: Dates and moments when the financial statements are disclosed.
(5) *Time Ontology*: An ontology of temporal concepts used to represent facts about topological relations among instants and time intervals and information about durations and date time [21, 22]. In the mixed semantic model, the Time Ontology vocabulary supports the publication of financial reporting periods in the form of linked data.
(6) *Temporal Entity Taxonomy*: Reuses the *Time Ontology* vocabulary and provides metadata on the reporting periods. The temporal entity taxonomy includes four components:

- *Period*: The time span covered by a reported financial statement.
- *Start date*: Indicates when (month, day, and year) the reporting period begins.
- *End date*: Indicates when (month, date, and year) the reporting period ends.
- *Instant*: Indicates when the financial statements were disclosed, which indicates that such statements are valid in a given date and/or time.

(7) *Unit*: Represents information on the measuring units (e.g. dollars, shares, euros, or dollars per shares) used in the financial statements. That said, the units used in XBRL-based financial statements are established by the Units Registry. The goal of this registry is to standardize use of units to encourage consistency across digital financial reports [23].
(8) *Payments Ontology*: A set of general-purpose vocabulary used for representing organizational spending information. It is not specific to government or local government applications. In the mixed sematic model, the payments ontology vocabulary is reutilized in the Financial Unit Taxonomy, because it

allows organizational spending data to be represented in the form of linked data, perhaps as an experiment or to develop reusable tools and processes [24].

(9) **Financial Unit Taxonomy**: All the values reported in XBRL-based statements have a unit of measurement assigned [12]. The Financial Unit Taxonomy provides metadata on the units of measurement used in the financial statements.

- **Balance Sheet Unit**: This component inherits from the Unit information on the type of currency assigned to each value reported in a financial statement.

- **PerShare Item**: Related to the Balance Sheet Unit. Used for concepts measured in dollars per shares. The Unit reference in the instance file, in this case the balance sheet, must include the currency as a numerator and the quota as the denominator. In this sense, the measurement unit can be read as *USD* per share or *Rupee per share*, for instance.

- **Currency**: Also related to the Balance Sheet Unit and used for financial concepts expressed as amounts of money in a given monetary unit. This measurement unit is regulated by the International Standard for Currency Codes—ISO 4217 [25]. Two examples of data expressed under this standard are the American dollar and the Swiss franc. Following the ISO-3166 standard, the former is represented as USD, being US the country code and D the initial letter of the monetary unit, the dollar. On the other hand, according to the same standard, the Swiss franc is represented by CHF, where CH is Switzerland's country code, and F is the initial letter of the currency, the franc [26].

(10) **XBRL Statements Taxonomy**: This component represents the reutilization of financial taxonomies. Such taxonomies are structured in XML schemas and linkbases that define the financial statement's structure through the following linkbases: labels, references, presentation, calculation, and definition. Figure 3 depicts an RDF US-GAAP financial taxonomy graph segment for balance sheets.

The balance sheet taxonomy helps FINALGRANT calculate the financial ratios presented in Table 1 (see Sect. 2). Figure 4 introduces FINALGRANT's main graphical user interface (GUI).

4 Case Study: Financial Analysis of Walmart with FINALGRANT

In this case study, FINALGRANT is used to analyze the financial situation of Walmart Stores Inc. in 2012, 2013, and the first three quarters of 2014; namely from January 1, 2012 to November 11, 2014. The goal of this study is to demonstrate that FINALGRANT can be successfully used by a banking institution to analyze

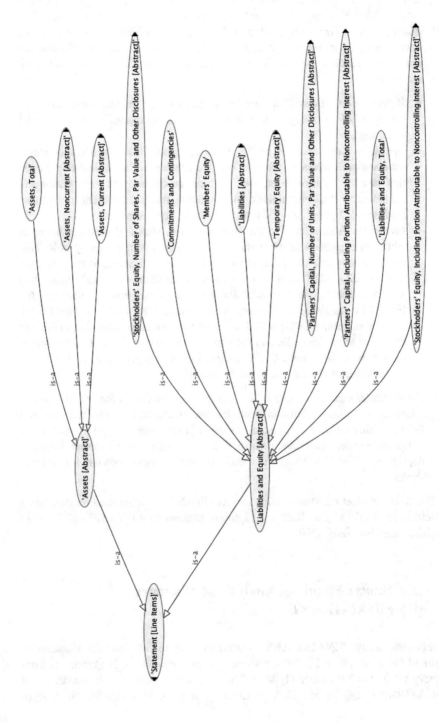

Fig. 3 RDF graph segment of the US-GAAP financial taxonomy for balance sheets

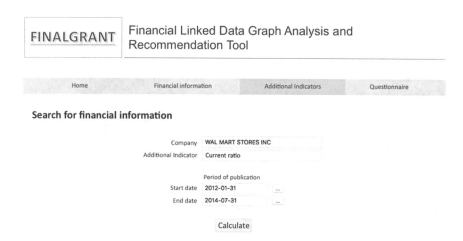

Fig. 4 Main graphical user interface of FINALGRANT

Walmart's financial situation and present the analysis result to a customer that might be willing to invest in the company's stock. That said, the study case situation can be presented as follows:

1. Mrs. Yannet Gómez Lara has 200,000.00 USD in savings and is thinking about investing this money in Walmart stock. However, Mrs. Gómez is not entirely sure of her decision, which is why she seeks the assistance of the Investment Funds services at BBVA Bancomer, a banking institution.
2. The financial assistants at the bank use FINALGRANT to analyze Walmart's financial ratios, present the results to Mrs. Gómez, and thus guide her in her decision.

In this scenario, FINALGRANT is used to calculate Walmart's liquidity ratios and debt ratio for the full years 2012 and 2013 and the first three quarters of 2014 (i.e. from January 1, 2012 to November 11, 2014). On the one hand, the liquidity ratios allow the bank and Mrs. Gómez to know Walmart's ability to turn its assets into cash and pay its short-term debt obligations. The Walmart liquidity ratios calculated include current ratio, working capital ratio, and acid test ratio. On the other hand, Walmart's debt ratio shows the proportion of the company's assets that are financed by debt.

4.1 Walmart's Current Ratio

The current ratio is a liquidity ratio of great interest, since it reflects a business ability to pay its debt obligations. The higher the current ratio, the more liquid a company, and thus the more capable it is to pay its debt. In this sense, a high current ratio is a warranty to shareholders, as it guarantees them a safe investment [8, 9].

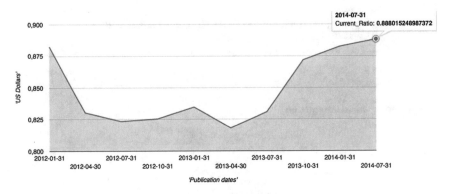

Fig. 5 Walmart's current ratio quarterly graph

*In million dollars (USD)

Current_Assets	Current_Liabilities	Current_Ratio_Value	Publication_Date
54975000000	62300000000	0.882423756019262	2012-01-31
57276000000	68978000000	0.830351706341152	2012-04-30
56259000000	68315000000	0.823523384322623	2012-07-31
63431000000	76845000000	0.825440822434771	2012-10-31
59940000000	71818000000	0.834609707872678	2013-01-31
60176000000	73552000000	0.818142266695671	2013-04-30
60002000000	72214000000	0.830891516880383	2013-07-31
67142000000	77021000000	0.871736279715922	2013-10-31
61185000000	69345000000	0.882327492969933	2014-01-31
59632000000	67152000000	0.888015248987372	2014-07-31

Fig. 6 Compendium of Walmart's quarterly current ratio

After the bank assistants enter the calculation parameters, FINALGRANT shows Walmart's current ratio for each quarter of the given period and calculates and shows the graph. In this sense, Fig. 5 shows that from January 1, 2012 to November 11, 2014, Walmart showed the highest current ratio in the third quarter of 2014. However, the company's solvency level was low, since for every dollar that Walmart had in short-tem debt, it could pay only 0.88 dollars, and 12 cents were still owed.

FINALGRANT provides a compendium of data to easily analyze, calculate, and interpret Walmart's current ratio. The data in this compendium are shown in Fig. 6 and correspond to the graph presented in Fig. 5 (Table 2).

Interpretation: The rules ((Current Liabilities * 100)/Current Assets) and ((Current Assets − Current Liabilities) * 100) applied to the annual data are used respectively to calculate the percentages of debt obligation payments and capital available. In this sense, Table 4 indicates that in 2012, for every dollar of current obligations (debt), Walmart had 3.36 to pay them. This means that 83% of Walmart's current assets that year were destined to pay the company's debt, and

Table 2 Walmart's annual current ratio

Walmart—Current ratio: 01/01/2012–11/11/2014					
Year	Current assets	Current liabilities	Current ratio (USD)	Obligations (%)	Capital available (%)
2012	231,941,000,000.00	276,438,000,000.00	3.36	83.90	16.10
2013	247,260,000,000.00	294,605,000,000.00	3.35	83.93	16.07
2014	120,817,000,000.00	136,497,000,000.00	1.76	88.51	11.49

only 16.10% were capital available. As for 2013, FINALGRANT shows that Walmart's liquidity declined, its debt-obligation payments increased up to 83.93%, and the company's available capital diminished (16.07%). Finally, in the third quarter of 2014, Walmart's current ratio did not improve in comparison with the previous two years.

4.2 Walmart's Working Capital Ratio

The working capital ratio seeks to guarantee a company's operational performance. A negative capital ratio implies that a company cannot cover its debt properly. On the other hand, a higher than 1 working capital ratio indicates that a company's assets exceed its liabilities and encourages financing and investments from shareholders [8, 9]. Figure 7 presents the data retrieved by FINALGRANT to calculate Walmart's working capital.

The results presented above suggest that for the analyzed period, Walmart lacked enough liquidity to cover its short-term debt. Yet this does not mean that the company went bankrupt or that it stopped paying its debts.

*In million dollars (USD)

Current_Assets	Current_Liabilities	Working_Capital_Value	Publication_Date
54975000000	62300000000	-7325000000	2012-01-31
57276000000	68978000000	-11702000000	2012-04-30
56259000000	68315000000	-12056000000	2012-07-31
63431000000	76845000000	-13414000000	2012-10-31
59940000000	71818000000	-11878000000	2013-01-31
60176000000	73552000000	-13376000000	2013-04-30
60002000000	72214000000	-12212000000	2013-07-31
67142000000	77021000000	-9879000000	2013-10-31
61185000000	69345000000	-8160000000	2014-01-31
59632000000	67152000000	-7520000000	2014-07-31

Fig. 7 Walmart's quarterly working capital

Table 3 Walmart's annual working capital

Walmart—Working capital: 01/01/2012–11/11/2014			
Year	Current assets	Current liabilities	Working capital
2012	231,941,000,000.00	276,438,000,000.00	−44,497,000,000.00
2013	197,260,000,000.00	294,605,000,000.00	−47,345,000,000.00
2014	120,817,000,000.00	136,497,000,000.00	−15,680,000,000.00

To summarize Walmart's overall short-term financial health, Table 3 presents the retailer's annual working capital for the period January 1, 2012–November 11, 2014.

Interpretation: The negative results from Table 3 reflect Walmart's unstable liquidity but are not a synonym of bankruptcy or temporary insolvency. The negative working capital ratios reflect the company's funding needs from either creditors or investors to increase its current assets and working capital. This would allow Walmart to pay off its debt.

4.3   Walmart's Acid-Test Ratio

The acid-test ratio reveals a business ability to pay off its current liabilities but removes inventory from the equation. A less than 1 acid-test ratio means that a company lacks the liquid current assets to cover its current liabilities, whereas a higher than 1 ratio can reflect an excess of liquidity. That said, the optimal acid-taste ratio ranges from 0.5 and 1 [8, 9]. Figure 8 presents the data retrieved by FINALGRANT to calculate Walmart's acid test ratio.

Interpretation: As can be observed in Fig. 8, Walmart's quarterly acid-test ratio ranged from 0.21 to 0.22 in the analyzed period. Since these values are not optimal, they imply that from January 1, 2012 to November 11, 2014, Walmart could not have paid off its current obligations without having inventory liquidation.

*In million dollars (USD)

Current_Assets	Current_Liabilities	Inventory	AcidTest_Value	publication_Date
54975000000	62300000000	40714000000	0.228908507223114	2012-01-31
57276000000	68978000000	41284000000	0.231842036591377	2012-04-30
56259000000	68315000000	40558000000	0.229832394056942	2012-07-31
63431000000	76845000000	47487000000	0.207482594833756	2012-10-31
59940000000	71818000000	43803000000	0.224692973906263	2013-01-31
60176000000	73552000000	43138000000	0.231645638459865	2013-04-30
60002000000	72214000000	42793000000	0.238305591713518	2013-07-31
67142000000	77021000000	49673000000	0.226808273068384	2013-10-31
61185000000	69345000000	44858000000	0.235445958612733	2014-01-31
59632000000	67152000000	45451000000	0.21117762687634	2014-07-31

Fig. 8 Walmart's quarterly acid-test ratio data

Also, according to the latest acid-test ratio, the company had only 21% of liquid assets to cover its short-term debts. In other words, in the third quarter of 2014, Walmart's short-term debts were 79% higher than its liquid assets.

4.4 Debt Ratio

The debt ratio indicates the degree of participation from stakeholders in a company's assets and is expressed as a percentage. Usually, a 60% debt ratio is considered to be manageable [8, 9], as it means that for every 100.00 USD of assets, the company has 60.00 USD worth of liabilities. In other words, 60% of the business' total assets are financed by debt.

To calculate Walmart's debts ratio, the following formula needs to be applied:

$$\text{Debt Ratio} = ((\text{Total Liabilities}/\text{Total Assets}) * 100)$$

To use the debt ratio formula, first we need to calculate the total liabilities ratio and the total assets ratio as presented below:

(a) **Total Liabilities (Current Liabilities + Non-Current Liabilities)**

    a. *Current Liabilities.*
    b. *Long-Term Debt, Total Current Maturities.*

(b) **Total Assets (Current Assets + Fixed Assets)**

    a. *Current Assets.*
    b. *Accumulated Depreciation, Depletion, Amortization, Plant, and Equipment.*

FINALGRANT calculates Walmart's aforementioned ratios to then estimate the company's debt ratio. The following section describes the calculation of Walmart's debt ratio for the analyzed time period.

4.4.1 Walmart's Debt Ratio

Figure 9 presents quarterly financial data regarding Walmart's current liabilities. Similarly, Fig. 10 introduces Walmart's long-term debt, including total current maturities. Both long-term debt and current maturities form total liabilities. Finally, Figs. 11 and 12 respectively present Walmart's current assets and accumulated depreciation, depletion, amortization, plant, and equipment. The two indicators together form Walmart's total assets.

After calculating Walmart's total liabilities and total assets, the bank assistants use FINALGRANT to estimate the company's debt ratio. The results of this process are depicted in Table 4.

Fig. 9 Walmart current
liabilities

Value_Ratio	Publication_date
62300000000	2012-01-31
68978000000	2012-04-30
68315000000	2012-07-31
76845000000	2012-10-31
71818000000	2013-01-31
73552000000	2013-04-30
72214000000	2013-07-31
77021000000	2013-10-31
69345000000	2014-01-31
67152000000	2014-07-31

Fig. 10 Walmart's long-term
debt

Value_Ratio	Publication_date
1975000000	2012-01-31
2509000000	2012-04-30
4029000000	2012-07-31
6550000000	2012-10-31
5587000000	2013-01-31
5967000000	2013-04-30
4692000000	2013-07-31
4147000000	2013-10-31
4103000000	2014-01-31
4659000000	2014-07-31

Fig. 11 Walmart current
assets

*Values in million dollars (US)

Value_Ratio	Publication_date
54975000000	2012-01-31
57276000000	2012-04-30
56259000000	2012-07-31
63431000000	2012-10-31
59940000000	2013-01-31
60176000000	2013-04-30
60002000000	2013-07-31
67142000000	2013-10-31
61185000000	2014-01-31
59632000000	2014-07-31

Fig. 12 Walmart's
accumulated depreciation,
depletion, and amortization,
plant, and equipment

*Values in million dollars (US)

Value_Ratio	Publication_date
45399000000	2012-01-31
47600000000	2012-04-30
48961000000	2012-07-31
50450000000	2012-10-31
51896000000	2013-01-31
53395000000	2013-04-30
54724000000	2013-07-31
56313000000	2013-10-31
57725000000	2014-01-31
61709000000	2014-07-31

Table 4 Walmart's annual debt ratio

Walmart—Debt ratio: 01/01/2012–11/11/2014			
Year	Total liabilities	Total assets	Debt ratio (%)
2012	291,501,000,000.00	424,351,000,000.00	68
2013	314,998,000,000.00	463,588,000,000.00	67
2014	145,259,000,000.00	240,251,000,000.00	60

Interpretation: The data reveal that in 2012 Walmart had 68% of its total assets financed by debt. This debt ratio of 68% is risky, as it indicates that for every 100 USD of assets, the company had 68.00 USD worth of debt. Walmart's debt ratio fell 1% in 2013, yet it was still risky. However, the debt ratio significantly improved in the third quarter of 2014, as it fell to 60%, which according to experts, is manageable.

4.5 Additional Data

After estimating the financial ratios, the bank assistants collect reference information on Walmart. Figure 13 shows a fragment of DBpedia interfaces accessed by FINALGRANT to collect such information.

The DBpedia links allow bank assistants to collect reference information on Walmart to provide Mrs. Gómez more accurate suggestions. Such information is obtained from the linked open data (LOD) cloud—i.e. DBpedia [27]—and refers to the company's founding history (founder and place), the number of employees, headquarter location, types of products offered, and additional reference links. Finally, after a process of data search, navigation, calculation, analysis, and interpretation using FINALGRANT, the bank assistants can offer Mrs. Gómez well-founded suggestions to either invest or not invest in Walmart stock. Such suggestions and conclusions are presented in the following section.

Fig. 13 Walmart DBpedia links through FINALGRANT

4.6 Case Study Conclusions

The results estimated by FINALGRANT with regards Walmart's financial situation from January 1, 2012 to November 11, 2014 can be presented by the bank assistants to Mrs. Gómez as the following conclusion:

1. Walmart is an American multinational retail corporation that highly depends on its inventory sales to cover its current short-term obligations (if they were to be required). So far, the company is a viable opportunity for Mrs. Gómez to invest.
2. Walmart's current ratio analysis suggests that the company is usually able to pay its debt, yet for 2012 and 2013, Walmart's current ratio was of 1.14 USD. Such results imply that the retailer has an acceptable but not optimal short-term liquidity and must increase its current ratio to at least 1.50 USD per year.
3. Walmart does not show favorable financial results in terms of its working capital. The company's working capital ratio was negative at any time (2012, 2013, and 2014), perhaps as a result of a poor investment plan. Consequently, the retailer needs borrowings from either investors or creditors.
4. The acid test analysis did not show optimal values in any of the years analyzed. Walmart's acid-test ratio ranged from 0.21 to 0.22, implying that from January 1, 2012 to November 11, 2014, Walmart could not have paid off its current obligations without selling its inventory.
5. The debt ratio analysis indicates that in 2012 and 2013 Walmart had a risky debt ratio—68 and 67%, respectively. However, this ratio significantly improved in the third quarter of 2014, as it fell to 60%, which according to experts, is manageable.

Following such conclusions, the bank assistants recommend Mrs. Gómez waiting for at least six more months to reanalyze Walmart's indicators and see whether they have improved or it is more viable to invest in a similar retail corporation, such as Costco Wholesale Corp. This recommendation seems objective, since a company must be able to both pay its debt and incur obligations.

5 Research Conclusions and Further Work Recommendations

This research discusses basic financial and accounting terms and concepts. Similarly, we explore XBRL as a standard for digital business reporting and propose FINALGRANT, a financial linked data graph analysis and recommendation tool that supports fund investment decisions. In the case study, FINALGRANT allows the bank assistants to visualize, calculate, and analyze four financial indicators for Walmart Stores Inc., an American multinational retailer corporation. Such data are used to suggest, Mrs. Gómez, a bank customer, that she should not invest in the retailer's stock and should look for another alternative.

As further research, FINALGRANT can be provided with automated recommendation capabilities by adding the necessary heuristic rules. Also, further research can analyze the financial situation of non-American corporations, yet EDGAR is so far the largest repository of XBRL-based digital financial statements, which makes it the most viable data source for FINALGRANT to transform XBRL data into linked data. That said, other organizations, such as the National Securities Market Commission in Spain are following the trend of digital XBRL-based business reporting and can eventually represent potential sources of data to FINALGRANT.

Finally, we recommend applying sentiment analysis algorithms to FINALGRANT to incorporate social media data, namely user opinions, regarding the analyzed companies. The results obtained can be interlinked with the financial data retrieved by FINALGRANT to determine whether social media opinions match or influence on the financial situation of a corporation.

Acknowledgements The authors are grateful to the National Technological Institute of Mexico for supporting this work. This research was also sponsored by the National Council of Science and Technology (CONACYT) and the Secretariat of Public Education (SEP) through PRODEP.

References

1. Hoffman, C., Van-Egmond, R.: Digital Financial Reporting Using an XBRL-Based Model (2012)
2. Bray, T., Paoli, J., Sperberg-McQueen, C.M., Maler, E., Yergeau, F.: Extensible markup language (XML). World Wide Web Consort. Recomm. REC-xml-19980210. http://www.w3.org/TR/1998/REC-xml-19980210 (1998)
3. Williams, J.R., Haka, S.F., Bettner, M.S., Carcello, J.V: Financial and Managerial Accounting. China Machine Press (2005)
4. Zhu, H., Madnick, S.E.: Semantic integration approach to efficient business data supply chain: integration approach to interoperable XBRL. MIT Sloan School of Management Research Paper Series (2007)
5. Lev, B., Thiagarajan, S.R.: Fundamental information analysis. J. Account. Res. **31**, 190–215 (1993)
6. Delen, D., Kuzey, C., Uyar, A.: Measuring firm performance using financial ratios: a decision tree approach. Expert Syst. Appl. **40**, 3970–3983 (2013)
7. Bliss, J.H.: Financial and operating ratios in management. The Ronald Press Company (1923)
8. Montero, J.M., Fernández-Aviles, G.: Enciclopedia de economía, finanzas y negocios. Editorial CISS (Grupo Wolters Kluwer), Madrid (2010)
9. Kimmel, P.D., Weygandt, J.J., Kieso, D.E.: Financial accounting: tools for business decision making. Wiley (2010)
10. U.S. SEC: Filer Manual—Volume II EDGAR Filing (2013)
11. Penman, S.H.S.: Financial statement analysis and security valuation. McGraw-Hill Education, New York (2009)
12. XBRL-España: Libro blanco XBRL (2006)
13. Thompson, H.: XML Schema Part 1: Structures, 2nd edn (2012)
14. Biron, P., Malhotra, A.: W3C-Group: XML Schema Part 2: Datatypes
15. Engel, P., Hamscher, W., Advantage, S., Shuetrim, G., vun Kannon, D., Pryde, C.: Extensible Business Reporting Language (XBRL) 2.1 July 2, pp. 1–165 (2008)

16. Lindörfer, F.: Semantic Web Frameworks-Jena, Joseki, Fuseki & Pellet (2010)
17. Erling, O., Mikhailov, I.: RDF support in the virtuoso DBMS. In: Networked Knowledge-Networked Media, pp. 7–24. Springer, Berlin (2009)
18. Hausenblas, M.: Exploiting linked data to build web applications. IEEE Internet Comput. **13**, 68–73 (2009)
19. Berners-Lee, T.: Linked Data—Design Issues
20. Radzimski, M., Sánchez-Cervantes, J.L., Garcia-Crespo, A., Temiño-Aguirre, I.: Intelligent architecture for comparative analysis of public companies using semantics and XBRL data. Int. J. Softw. Eng. Knowl. Eng. **24**, 801–823 (2014)
21. Hobbs, J.R., Pan, F.: An ontology of time for the semantic web. **3**, 66–85 (2004)
22. Hobbs, J.R., Pan, F.: Time Ontology in OWL. W3C Working Draft, 27 Sept 2006
23. Pryde, C., Piechocki, M., John, C. St., Warren, P., North, D.: Units Registry—Structure 1.0
24. Reynolds, D.: Payments Ontology: Reference
25. ISO: Currency Codes—ISO 4217
26. ISO: Country Codes—ISO 3166
27. Auer, S., Bizer, C., Kobilarov, G., Lehmann, J., Cyganiak, R., Ives, Z.: DBpedia: a nucleus for a web of open data. In: Aberer, K., Choi, K.-S., Noy, N., Allemang, D., Lee, K.-I., Nixon, L., Golbeck, J., Mika, P., Maynard, D., Mizoguchi, R., Schreiber, G., Cudré-Mauroux, P. (eds.) The semantic web SE—52, pp. 722–735. Springer, Berlin (2007)

Constructing and Interrogating Actor Histories

Tony Clark, Vinay Kulkarni, Souvik Barat and Balbir Barn

Abstract Complex systems, such as organizations, can be represented as executable simulation models using actor-based languages. Decision-making can be supported by system simulation so that different configurations provide a basis for *what-if* analysis. Actor-based models are expressed in terms of large numbers of concurrent actors that communicate using asynchronous messages leading to complex non-deterministic behaviour. This chapter addresses the problem of analyzing the results of model executions and proposes a general approach that can be added to any actor-based system. The approach uses a logic programming language with temporal extensions to query execution traces. The approach has been implemented and is shown to support a representative system model.

1 Introduction

Organizations and systems can be simulated using Multi-Agent Systems [1–3]. This approach builds a model of an organisation in terms of independent goal-directed agents that concurrently engage in tasks, both independently and collaboratively. Collections of such agents form an executable model that produces results. Fishwick [4] notes the key features of computer simulation to be modelling, execution and analysis of output. An important reason for using agents for simulation is

T. Clark (✉)
Sheffield Hallam University, Sheffield, UK
e-mail: t.clark@shu.ac.uk

V. Kulkarni · S. Barat
Tata Consultancy Services Research, Pune, India
e-mail: vinay.vkulkarni@tcs.com

S. Barat
e-mail: souvik.barat@tcs.com

B. Barn
Middlesex University, London, UK
e-mail: b.barn@mdx.ac.uk

© Springer International Publishing AG 2018
R. Valencia-García et al. (eds.), *Exploring Intelligent Decision Support Systems*,
Studies in Computational Intelligence 764,
https://doi.org/10.1007/978-3-319-74002-7_2

that the systems of interest are complex, for example because they involve socio-technical features [5]. As noted in [6]: *humans use patterns to order the world and make sense of things in complex situations*, and it follows that pattern-based analysis may be used to analyze an agent-based simulation model. This chapter addresses the problem of how to create and analyze agent-based simulations.

Our work on simulation for decision support [7–12] has led to the design of a simulation workbench built around an actor language [13] called ESL [14]. The language ESL is used to construct agent-based simulation models that are run to produce histories. Each history contains a sequence of events produced by the behaviour of the actors in the simulation and thereby captures their emergent behaviour. A history is a temporal database of facts describing the states of, and communications between, actors in the simulation.

The research question that we seek to investigate is: *what general-purpose mechanism can be devised to generate actor histories and then analyze them using temporal queries? Where possible the mechanism should be applicable to existing actor-based technologies and use standard query languages with minimal extensions*. We take a design-based approach to this research by taking an existing technology and implementing extensions that support the production and analysis of actor-histories.

The current state of the practice of analysis of simulation results is predominantly based on the visualization and human interpretation. We propose a programmatic approach to the construction and interrogation of simulation histories. History construction is achieved by extending the standard operational actor model of computation [15, 16] in order to capture temporal events during simulation execution. History interrogation is achieved by extending standard logic programming with temporal operators that are defined in terms of a supplied history containing time-stamped events.

Our contribution is a pair of general-purpose languages for the construction and subsequent interrogation of agent-based execution histories. In both cases conventional computational models are extended with novel mechanisms for histories: an interpreter for actor languages is extended with primitives for history production and a Prolog meta-interpreter is extended to support history interrogation.

The proposed approach is evaluated in terms of its completeness, viability and validity. Completeness follows from the universality of the actor model of computation, from our claim that our actor interpreter generates all key computational events, and from our claim that the query language can express all queries of interest. Viability is demonstrated by our implementation of a simulation workbench and validity is demonstrated by showing how the implementation supports the construction and interrogation of a representative simulation. The conclusion discusses threats to validity, how we plan to address them, and outlines next steps.

2 Related Work

The use of Multi-Agent Systems (MAS) for system simulation has been explored by a number of researchers, for example in [17–20], where agent simulation models range from collection of numerical equations to sophisticated behaviours encoded using a BDI-based approach. Researchers have developed approaches for the definition and analysis of simulation properties. In [17], Bosse et al. present a generic language for the formal specification and analysis of dynamic properties of MAS that supports the specification of both qualitative and quantitative features, and therefore subsumes specification languages based on differential equations. However, this is not an executable language like that presented in this chapter. It has been specialized for simulation and has produced the LEADSTO language [21] that is a declarative order-sorted temporal language where time is described by real numbers and where properties are modelled as direct temporal dependencies between properties in successive states. Though quite useful in specifying simulations of dynamic systems, it does not provide any help in querying the resultant behaviour. Bosse et al. further propose a multi-agent model for mutual absorption of emotions to investigate emotion as a collective property of a group using simulation [22]. It provides mathematical machinery to validate a pre-defined property over simulation trace. However, there is no support for temporal logic operators.

Sukthankar and Sycara propose an algorithm to recognize team behaviour from spacio-temporal traces of individual agent behaviours using dynamic programming techniques [23], but do not address general behavioural properties arising from simulations. Vasconcelos et al. present mechanisms based on first-order unification and constraint solving techniques for the detection and resolution of normative conflicts concerning adoption and removal of permissions, obligations and prohibitions in societies of agents [24].

The tool described in [25] produces static diagrams of agent communication topologies using a *society* tool. The authors support off-line video-style replay facilities with forward and backward video modes as a powerful debugging aid. However there is no programmatic mechanism for interrogating the histories.

Temporal logics have been used to specify the behaviour of MAS [26] and to analyze the specification for properties using theorem proving or model checking. Our approach uses a similar collection of temporal operators, however we are applying the behaviour specifications post hoc in order to investigate whether a given behaviour took place, rather than to express required behaviour or to analyze properties such as consistency etc.

The need to analyze agent-based simulations is related to the field of agent-based system testing. As noted in [27] *attempting to obtain assurance of a system's correctness by testing the system as a whole is not feasible* and *there is, at present, no practical way of assuring that they will behave appropriately in all possible situations.* Our approach is intended to be a pragmatic partial solution that is used selectively in collaboration with a domain expert. Queries can be used to test whether properties exist in particular histories, and could help scope the use of more formal static methods.

Using temporal operators to construct queries over databases is a standard approach. Queries can be encoded in logic [28] or in SQL extensions [29], although as noted in [30]: *Much of real-life data is temporal in nature, and there is an increasing application demand for temporal models and operations in databases. Nevertheless, SQL:2011 has only recently overcome a decade-long standstill on standardizing temporal features. As a result, few database systems provide any temporal support, and even those only have limited expressiveness and poor performance.* A logic provides increased expressive power over an SQL-like language at the cost of requiring a theorem prover or a model checker with the associated scalability issues. Our approach, using logic-programming, aims to be more expressive than SQL whilst addressing scalability.

Managing temporal data is becoming increasingly important for many applications [31, 32]. Our work is related to process mining from the event logs that are created by enterprise systems [33] where queries can be formulated in terms of a temporal logic and applied to data produced by monitoring real business systems. Other researchers have proposed adding temporal operators to query languages in order to process knowledge bases [34]. We have extended these approaches in the context of simulation histories by showing how to encode them in an operational query language.

The nature of agent-based systems leads to high levels of concurrency with an associated challenge regarding system analysis when behaviour is not as expected. As reported in [35]: *in order to locate the cause of such behaviour, it is essential to explain how and why it is generated* [36]. If histories are linked to source code then queries can be used as part of an interactive debugger, extending the approach described in [35]. In conclusion, there have been a number of approaches in the literature that analyze actor-based systems, some of them are based on histories, but none provide the expressive power of the language defined in this chapter. Furthermore, we show how any actor-based language can be extended to produce histories that are suitable for interrogation by queries written in an extended logic programming language.

3   Actors and Histories

An agent-based simulation model consists of agents, each of which has local knowledge, goals and behaviour. Such a model can be operationalized in terms of the actor model of computation whereby each actor has an independent thread of control, has a private state and communicates with other actors via asynchronous messages as shown in Fig. 1. For the purposes of this chapter we conflate the terms *agent* and *actor*.

Fig. 1 Actor model of
computation [39]

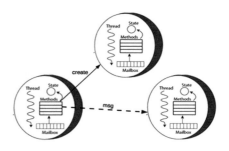

The key features of the actor computation model are [37–39]: (1) the creation of
new actors; (2) sending asynchronous messages; (3) updating a local variable;
(4) changing behaviour. The latter allows an actor to adapt by changing the way in
which it responds to messages. The state of an actor can be represented in terms of
its local variable storage (including references to other actors), its thread of execution, and its message queue. Execution proceeds independently at each actor by
selecting the next message on the queue, using the message to index a suitable
handler in the actor's behaviour description, and proceeding to execute the handler
on the actor's thread. When the execution terminates, it repeats the process by
selecting the next message.

Consider a situation where a customer processes jobs on a machine. The customer submits a job request to a machine that may subsequently result in a notification that the job has been completed, or that the machine is busy and cannot
accept the job. After accepting a job, a machine may break down causing a delay.
A simple actor model for this situation is shown in Fig. 2a where the types
Customer and Machine and the behaviours workingMachine and
brokenMachine. A type defines an interface that may be implemented by many
different behaviours, and a behaviour is equivalent to a Java class that can be
instantiated to produce actors. The behaviour of a machine is shown in Fig. 2b and
is distributed between the two behaviour definitions: a machine initially has the
behaviour workingMachine and is Idle. A working machine becomes Busy
when it receives a job request, and may change behaviour to become a
brokenMachine. When broken, the Machine interface is implemented differently and may become working after a period of repair.

Figure 2c shows an execution history corresponding to the machine and customer actors. The simulation is driven by messages Time(n) which are generated
at regular intervals, and the history contains the events that are produced at each
time interval. The event types are: New(b,i) where i is a unique actor identifier,
and b is the corresponding behaviour; Update(i,n,v) where i is an actor
identifier, n is the name of a state variable, and v is a new value for the variable;
Send(s,t,m) records a message m being sent from actor s to target t; Consume
(i,m) removes message m from the head of the message queue for actor i;
Become(i,b) records the change of behaviour of actor i to have behaviour b.

The machine example demonstrates typical features of actor-based simulation:
time and stochastic behaviour. All actors receive Time(n) messages that drive the

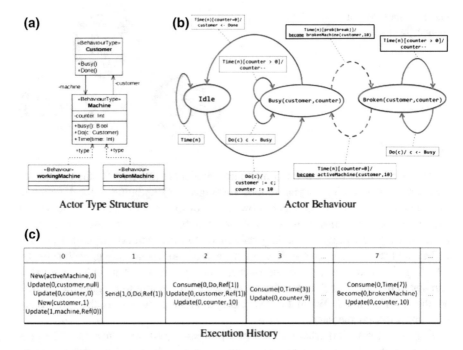

Fig. 2 Features of actor behaviour

simulation and therefore events can be associated with a specific time, thereby providing an event ordering within a specific history. The event that causes a machine breakdown is dependent on a given probability: `break`, and therefore multiple runs of the same actor model can produce different histories corresponding to emergent behaviour. Analysis of a history is based on detecting patterns in the sequence of events.

4   Constructing Histories

The actor model of computation has been implemented in a significant number of languages and libraries [40]. The implementations differ in terms of syntax and in the integration with other language features, however the key aspects of the actor model are common to all. This section shows how any of these implementations can be extended to produce histories by defining an actor interpreter that abstracts away all non-essential features and that has been minimally extended to produce histories. The interpreter is defined using a functional subset of ESL, which supports basic data types, simple algebraic data, lists and sets. The latter is used in conjunction

```
1       type Id       = Int;  // Actor identifiers.
2       type Time     = Int;  // Atomic time units.
3       type Behaviour = [Handler(Str,[Command])];  // Handlers.
4       data Command  =      // Abstract commands.
5          | Send(Id,Str)   // Send a name to a target actor.
6          | Update(Str)    // Change a variable.
7          | Block([Command])  // Group of commands.
8          | Become(Str)    // Change behaviour.
9          | New(Str);      // Create a new actor.
10         | Choice(Set(Command))  // Choice between alternatives.
11         | Consume        // Handle next message.
12      type Queue    = [Message(Str)];  // Sequence of messages.
13      type Actor    = Machine(Id,     // Actor's unique id.
14         | [Command],     // Instructions on actor thread.
15         | Behaviour,     // Actor message handlers.
16         | Queue,         // Pending messages.
17         | Time);         // Computation steps available.
18      type DB = [Fact(Time,Command)];     // A History of facts.
19      type ESL = State(Set{Actor},DB,Time);  // The system state.
```

Fig. 3 An abstract model of actor programs as an ESL data type

with pattern matching to support non-deterministic set element selection, which is key to the *fairness* property of actor systems.

Figure 3 shows a model of actor program states: State(a,db,t) is the state of an executing actor system where a is a set of actors, db is a history database, and t is the current time. We are interested in specifying how db is constructed through the conventional operational semantics of actors. This is achieved by defining a single-step operational semantics: s2=step(s1) where system state s1 performs an execution step in order to become state s2. The complete execution of a system can be constructed by repeated application of step.

It is useful in simulations to be able to refer to global time via a clock. This can be used to schedule future computation or to allow actors to perform joint actions. To support the notion of global time, each actor in our operational model receives a regular *Time* message where each global time unit is measured in machine instructions. This mechanism seems to be fair and, although is not related to real-time, provides a basis for time that is useful in a simulation. To support this, each actor has an instruction count that, when reached, halts the actor. When all actors have been halted, global time is increased, and a message is sent to all actors.

An actor is represented as Machine(i,cs,b,q,t) where i is a unique identifier, cs are machine instructions that are currently executing on the actor's thread of control, b is the actor's behaviour, q is a message queue, and t is an integer that represents the number of instructions left to this actor within this global time unit. The function step is defined:

```
1    is0::(Actor) -> Bool
2    is0(Machine(i,cs,b,ms,t))          = t=0;
3
4    all0::(Set{Actor}) -> Bool
5    all0(set{})                        = true;
6    all0(set{a | as})                  = is0(a) and all0(as);
7
8    tickActor::(Actor) -> Actor
9    tickActor(Machine(i,cs,b,q,t))     = Machine(i,cs,b,Message('Time'):q,100);
10
11   tick::(Set{Actor}) -> Set{Actor}
12   tick(set{})                        = set{};
13   tick(set{a | s})                   = set{tickActor(a) | tick(s)};
14
15   step::(ESL) -> ESL
16   step(State(a,db,t)) when all0(a)   = State(tick(a),db,t+1);
17   step(State(set{m | ms},db,t))      = newState(m,ms,db,t)
```

where line 6 detects the situation where all actors have exhausted their execution resources for the current time unit and line 17 non-deterministically selects an actor m such that newState performs an execution step for m if that is possible.

Figure 4 defines function newState that performs a single step of execution for an actor with respect to the system. It is defined by case analysis on the actor's control instructions. In the case that the actor has exhausted its control and has no further messages (line 10) it can do nothing. If the control is exhausted and there is a pending message (line 11) then the control is updated to become the new message handler. Otherwise, lines 13–34 show how each command causes the history database to be extended with a fact that is labelled with the current global time. The operation newActor::(Str)->(Behaviour,Id) is used to create a new actor when supplied with the name of a behaviour. It is not defined in Fig. 4 but assumes a global collection of behaviour definitions and allocates a new actor identifier each time it is called.

A history database db is created from an initial configuration of actors a by repeated application of step until a terminal state is achieved such that all actors are exhausted and have no pending messages, db=getDB(run(a)):

```
1     send::(Int,Str,Set{Actor}) -> Set{Actor}
2     send(a,n,set{Machine(i,cs,b,q,t) I ms}) when i=a =
3     set{Machine(i,cs,b,Message(n):q,t) I ms}
4
5     getHandler::(Str,Behaviour) -> [Command]
6     getHandler(n,[])        = [];
7     getHandler(n,Handler(m,cs):b) = if m=n then cs else getHandler(n,b);
8
9     newState::(Actor,Set{Actor},DB,Time) -> ESL
10    newState(Machine(id,[],b,[],ta),ms,db,t) = State(set{mIms},db,t);
11    newState(Machine(id,[],b,Message(n):q,ta),ms,db,t) =
12      State(set{Machine(id,getHandler(n,b),b,q,ta-1) I ms},Fact(t,Consume):db,t);
13    newState(Machine(id,Send(target,n):cs,b,q,ta),ms,db,t) =
14      State(set{Machine(id,cs,b,q,ta-1) I send(target,n,ms)},
15           Fact(t,Send(target,n)):db,
16           t);
17    newState(Machine(id,Update(var):cs,b,q,ta),ms,db,t) =
18      State(set{Machine(id,cs,b,q,ta-1) I ms},
19           Fact(t,Update(var)):db,
20           t);
21    newState(Machine(id,Become(var):cs,b,q,ta),ms,db,t) =
22      State(set{Machine(id,cs,getBehaviour(var),q,ta-1) I ms},
23           Fact(t,Become(var)):db,
24           t);
25    newState(Machine(id,New(var):cs,b,q,ta),ms,db,t) =
26      State(set{Machine(id,cs,b,q,ta),Machine(id',[],b',[],100) I ms},
27           Fact(t,New(var))):db,
28           t) where (b',id') = newActor(var);
29    newState(Machine(id,Block(commands):cs,b,q,ta),ms,db,t) =
30      State(set{Machine(id,commands + cs,b,q,ta-1) I ms},db,t);
31    newState(set{Machine(id,Choice(set{cl_}):cs,b,q,ta) I ms},db,t) =
32    newState(set{Machine(id,c:cs,b,q,ta) I ms},db,t)
33    newState(m=Machine(id,Consume:cs,b,q,ta),ms,db,t) = State(set{m I ms},db,t)
```

Fig. 4 Semantics of ESL

act **workingMachine**(customer::Customer,counter::**Int**)::Machine {
  customer::Customer = **null**;
  Do(c::Customer) **when** counter > 0 -> c <- Busy;
  Do(c::Customer) **when** counter = 0 -> { counter := 10; customer := c }
  Time(n::**Int**) **when** customer <> **null and** counter > 0 ->
    probably(break) **become** brokenMachine(customer,10)
    else counter := counter - 1
  Time(n::**Int**) **when** customer <> null **and** counter = 0 -> c <- Done
}

act brokenMachine(customer::Customer,counter::**Int**)::Machine {
  Do(c::Customer)   -> c <- Busy;
  Time(n::**Int**) **when** counter = 0 -> **become** workingMachine(customer,10);
  Time(n::**Int**)   -> counter := counter - 1
}

Fig. 5 ESL implementation of machine

```
isTerminal::(ESL) -> Bool;
isTerminal(State(a,_,_))           = allTerminated(a);

allTerminated::(Set{Actor}) -> Bool;
allTerminated(set{})               = true;
allTerminated(set{a | as})         = isTerminated(a) and allTerminated(as);

isTerminated::(Actor) -> Bool;
isTerminated(Machine(_,[],_,[],_)) = true;
isTerminated(_)                    = false;

getDB::(ESL) -> DB;
getDB(State(_,db,_))               = db;

initState::(Set{Actor}) -> ESL;
initState(a)                       = State(a,[],0)

run::(Set{Actor}) -> ESL;
run(a)                             = repeat(step,initState(a),isTerminal);
```

Figure 5 shows a concrete ESL implementation of the machine from Fig. 2. The abstract implementation, using the language defined in Fig. 3 is as follows:

```
behaviour workingMachine {
          -> block(update(customer),update(counter))
   Do(c)  -> choice(send(c,Busy),block(update(counter),update(customer)))
   Time   -> choice(become(brokenMachine),update(counter),send(c,Done))
}
behaviour brokenMachine {
          -> block(update(customer),update(counter))
   Do(c)  -> send(c,Busy)
   Time   -> choice(become(workingMachine),update(counter))
}
```

This section has described an abstract operational model for the construction of actor histories. A history is a collection of facts of the form `Fact(t, f)` where t is a timestamp and f is a term representing an actor execution step. The semantics is defined as an interpreter for an abstract actor language that can be used as the basis of designing a similar modification to a wide range of concrete languages and the relationship of the abstract language to ESL has been demonstrated. The next section shows how the histories produced by the interpreter can be interrogated using queries that are expressed using logic programming.

5 Interrogation of Histories

Simulations consist of many autonomous agents with independent behaviour and motivation. Consequently, the system behaviour is difficult to predict. Furthermore, the highly concurrent nature of the actor model of computation makes the simulation difficult to instrument in order to detect situations of interest. Therefore, we propose the construction of simulation histories as a suitable approach to simulation interrogation. Given such a history we would like to construct queries that determine whether particular relationships exist, where the relationships are defined in terms of the key features of actor computation. Logic programming, as exemplified by Prolog, would seem to be an ideal candidate for the construction of such queries, however standard Prolog does not provide intrinsic support for expressing the temporal features of such histories. We define a typed logic programming language and define an extended Prolog meta-interpreter with history interrogation features.

5.1 Typed Logic Programming

A basis for the history query language is a statically typed version of Prolog:

```
append[T]::([T],[T],[T]);
append[T]([],l,l) <- !;
append[T]([x|l1],l2,[x|l3]) <-    append[T](l1,l2,l3);

length[T]::([T],Int);
length[T]([],0) <- !;
length[T]([h|t],n) <- length[T](t,m), n := m + 1;

member[T]::(T,[T]);
member[T](x,[x|_]);
member[T](x,[_|l]) <-    member[T](x,l);

subset[T]::([T],[T]);
subset[T]([],[]);
subset[T]([x|l],[x|s]) <-    subset[T](l,s);
subset[T](l,[_|s]) <- subset[T](l,s);

lookup[V]::(Str,V,[Bind(Str,V)]);
lookup[V](n,v,[Bind(n,v) | _]);
lookup[V](n,v,[_|env]) <-lookup[V](n,v,env);
```

The examples above are standard Prolog rules that have been elaborated with static type information that is checked by the ESL Workbench before execution. The rules length and member use parametric polymorphism over the type T of elements in a list. The rule lookup is parametric with respect to the type of the bindings in the environment list.

Standard Prolog, as shown above, does not provide support for history interrogation. Histories are temporally ordered facts, so history interrogation will involve queries that need to express ordering relationships between, what are otherwise, standard Prolog facts. This suggests that adding temporal operators to Prolog [41] and integrating the history facts with a Prolog rule database will provide a suitable basis for interrogation. During execution, a query is at a particular time unit in the history and can match against any of the facts at that time in addition to matching against rules in the rule-base. Operators can be used to move forwards and backwards in time to adjust the portion of the history that is used to establish facts. Quantification over the time variable to allow queries such as: *fact f exists at some point in the history from this point*, and *fact f exists at all points in the history before this point* to provide a suitably expressive basis for defining history interrogations with a logic programming framework. The rest of this section defines such a mechanism.

5.2 Meta Representation

This section defines a data representation for logic-programming rules where the rule-body elements support temporal operators over histories. The data type Body describes the elements that can occur in a rule body as depicted in Fig. 6. The terms Call and Is represent standard Prolog body elements; all other elements are

```
data Value =            // Values occurring in rules.
  Term(Str,[Value])     // A term is a named sequence of values.
| Var(Str)              // A named logic variable.
| I(Int)                // An integer.
| S(Str);               // A string.

data Body =             // An element in the body of a rule.
  Call(Str,[Value])     // Call a rule, supplying values.
| Is(Str,Value)         // Unify a variable with a value.
| Start                 // The start of the history.
| End                   // The end of the history.
| Next([Body])          // Move forward one unit of time.
| Prev([Body])          // Move back one unit of time.
| Always([Body])        // Body is true from now on.
| Eventually([Body])    // Body is true at some time from now.
| Past([Body])          // Body is true at some time previously.
| Forall([Body],Value,Value);// All ways in which body is true.
```

Fig. 6 Data type definition

extensions to standard Prolog. The extensions all relate to *current time* that is used to index the time-stamp associated with the facts in the history:

```
type Time  = Int;
type Entry = Fact(Time,Str,[Value]);
type DB    = [Entry];
```

Elements Start and End are satisfied when the current time is *0* and the end of the history respectively. An element Next(es) is satisfied when the elements *es* are satisfied at now +1, similarly Prev(es) at now -1. An element Always (es) is satisfied when the elements *es* are satisfied at all times from now, similarly Past(es) all times before now. Element Eventually(es) is satisfied when es are satisfied at some time in the future.

5.3 Meta Interpreter

Figure 7 defines a meta-interpreter for the history query language. Given a query q (v1,...,vn), a program prog, a database db and a history end time t, the query is satisfied when call(0,t,db,'q',[v1,...,vn],prog) is satisfied with respect to the definitions given in the program.

The meta-interpreter is based on a standard operational semantics for Prolog that is extended with features to process the supplied database (the definition of Forall is omitted, but is consistent with standard Prolog). The rule call is used to process a body element of the form Call(n,vs) where *n* is the name of a fact and vs are the arguments. Conventional Prolog processes such a call using the definition of call defined on lines 8–13 where a rule named n with an appropriate arity is found in the program and is supplied with the argument values using `matchs`.

Figure 7 extends conventional Prolog rule calling by allowing the fact to be present in the history at *the current time* (line 8). Therefore, the facts in the history become added to the facts that can be conventionally deduced using the rules. The semantics of the additional types of body elements are processed by the *try* rule (lines 41–63) by modifying the value of the current time appropriately, For example, the rule for `Next` (lines 48–49) fails if the end of the history has been reached, otherwise it attempts to satisfy the elements `es` after incrementing the current time by *1*.

An example rule is `customers` where `customers(cs)` is satisfied when `cs` is a list of all the customer actor identifiers in the history:

```
1.   customers::([Int]);
2.   customers([]) <- end, !;
3.   customers(cs) <-
4.   forall[new(a,'customer')](a,cs'), next[customers(cs'')],
5.   append[Int](cs',cs'',cs);
```

Line 2 defines that there can be no customers if we are at the end of the history. Lines 3–5 define how to extract the customer identifiers from this point in the history: line 4 uses `forall` to match all database facts of the form `actor(a,'customer',_)` where this fact has been added to the database when a new customer actor is created. Then, `next` is used to advance the time so that `cs"` are all the customer actors from this point onwards.

6   Evaluation

The approach has been implemented as part of the ESL Workbench. ESL is an actor language that has been designed to support simulations. It has static types and compiles to run on a virtual machine (VM) implemented in Java. We have extended the ESL VM with features to produce histories and then integrated a query language, implemented as a Prolog VM, extended with features to process histories. This section evaluates the validity of the approach by applying it to a case study. ESL has been used to construct a number of simulations including an IT service provider, a research institute, and the effect of the 2016 Indian Demonetisation initiative. In this chapter we use a case study that is based on existing work on agent-based organisation simulations [42] involving a shop where customers browse for items, seek help from assistants, and queue to buy chosen items. Customers become unhappy if they wait too long for help or in a queue, and unhappy customers leave the shop.

The shop would like to simulate customer and assistant behavior in order to minimize unhappy customers. The case study will be used to demonstrate the

Fig. 7 Query language meta
interpreter

```
1.  type Prog = [Rule(Str,[Value],[Body])];
2.  type Env = [Bind(Str,Value)];
3.
4.  rule::(Str,Rule(Str,[Value],[Body]),Prog);
5.  rule(n,Rule(n,as,body),[Rule(n,as,body)|prog]);
6.  rule(n,r,[_|prog]) <- rule(n,r,prog);
7.
8.  call ::(Time,Time,DB,Str,[Value],Prog);
9.  call(time,eot,db,n,vs,prog) <- member[Entry](Fact(time,n,vs),db);
10.  call(time,eot,db,n,vs,prog) <-
11.      rule(n,Rule(n,as,body),prog),
12.      length[Value](vs,l), length[Value](as,l),
13.      matchs(as,vs,[],vars), trys(time,eot,db,body,vars,_,prog);
14.
15.  eval::(Value,Env,Value);
16.  eval(Term('+',[l,r]),e,I(i)) < eval(l,e,lv), eval(r,e,rv), i := lv + rv;
17.  eval(Var(n),env,v) <- lookup[Value](n,v,env);
18.  eval(I(i),env,I(i));
19.  eval(S(s),env,S(s));
20.
21.  matchs::([Value],[Value],Env,Env);
22.  matchs([],[],env,env);
23.  matchs([a|as],[v|vs],in,out) <- match(a,v,in,in'), matchs(as,vs,in',out);
24.  match::(Value,Value,Env,Env);
25.  match(Term(n,vs),Term(n,vs'),in,out) <- matchs(vs,vs',in,out);
26.  match(Var(n),v,e,e) <- lookup[Value](n,v,e);
27.  match(Var(n),v,env,[Bind(n,v) | env]);
28.  match(I(i),I(i),env,env);
29.  match(S(s),S(s),env,env);
30.
31.  derefs::([Value],[Value],Env,Env);
32.  derefs([],[],env,env);
33.  derefs([v|vs],[v'|vs'],in,out) <- deref(v,v',in,in'), derefs(vs,vs',in',out);
34.  deref::(Value,Value,Env,Env);
35.  deref(Term(n,vs),Term(n,vs'),in,out) <- derefs(vs,vs',in,out);
36.  deref(Var(n),v,e,e) <- lookup[Value](n,v,e),!;
37.  deref(Var(n),v,env,[Bind(n,v) | env]);
38.  deref(I(n),I(n),env,env);
39.  deref(S(s),S(s),env,env);
40.
41.  trys::(Time,Time,DB,[Body],Env,Env,Prog);
42.  trys(_,_,_,[],env,env,prog);
43.  trys(t,eot,db,[e|es],i,o,p) <- try(t,eot,db,e,i,j,p), trys(t,eot,db,es,j,o,p);
44.  try::(Time,Time,DB,Body,Env,Env,Prog);
45.  try(t,eot,db,Call(n,vs),i,o,p) <- derefs(vs,vs',i,o),call(t,eot,db,n,vs',p);
46.  try(eot,eot,db,End,env,env,prog);
47.  try(0,_,db,Start,env,env,prog);
48.  try(eot,eot,db,Next(es),env,env,prog) <- !,false;
49.  try(t,eot,db,Next(es),i,o,p) <- t' := t + 1; trys(t',eot,db,es,i,o,p);
50.  try(0,eot,db,Prev(es),env,env,prog) <- !, false;
51.  try(t,eot,db,Prev(es),i,o,p) <- t' := t - 1, trys(t',eot,db,es,i,o,p);
52.  try(eot,eot,db,Always(es),env,out,prog) <- !;
53.  try(time,eot,db,Always(es),in,out,prog) <-
54.      trys(time,eot,db,es,in,in',prog), time' := time + 1,
55.      try(time',eot,db,Always(es),in',out,prog);
56.  try(eot,eot,db,Eventually(es),in,out,prog) <- !, false;
57.  try(t,eot,db,Eventually(es),i,o,p) <- trys(t,eot,db,es,i,o,p);
58.  try(time,eot,db,Eventually(es),in,out,prog) <- time' := time + 1,
59.      try(time',eot,db,Eventually(es),in,out,prog);
60.  try(0,eot,db,Past(es),in,out,prog) <- !, false;
61.  try(time,eot,db,Past(es),in,out,prog) <- trys(time,eot,db,es,in,out,prog);
62.  try(t,eot,db,Past(es),i,o,p) <- t' := t - 1, try(t',eot,db,Past(es),i,o,p);
63.  try(t,eot,db,Is(n,exp),e,e,p) <- eval(exp,e,v), lookup[Value](n,v,e);
```

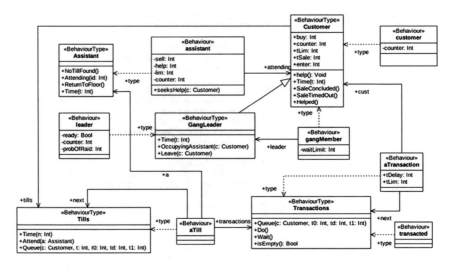

Fig. 8 Structure of shop actors

construction of an agent model that produces a history and the subsequent interrogation via a query. The query is chosen to demonstrate the utility of logic programming using the ESL query language and will also be analyzed in terms of its
efficiency based on the implementation described in the previous section.

The structure of the shop simulation is shown in Fig. 8. and its behaviour is
shown in Fig. 9 a snapshot of the output from ESL is shown in Fig. 10 where 10
customers are mostly unsatisfied (a) and mostly satisfied (b).

The simulation is driven by Time messages and all actors implement a Time
transition for all states; the empty transitions arising from Time are omitted. Time
is used in Fig. 9g to show how customers waiting at a till can time out and
ultimately leave the shop. The value supplied to a transaction is tLim which
determines how long a customer is prepared to wait without a sale being concluded.

Customers who leave the shop because they have waited too long to be serviced
at a till are deemed to be *unhappy*. The shop is interested in how to organize its
assistants, sales and floor-walking strategies in order to minimize unhappy customers. Figure 10 shows the ESL Workbench output from two different simulation
configurations. Figure 10a shows the result of 10 customers, 5 tills and 3 sales
assistants where roughly 75% of the customers are left unhappy. The number of
assistants has been increased to 5 in Fig. 10b where the situation is reversed. Note
that the simulation has many random elements and therefore each run is different,
but the two outputs characterize the relative differences.

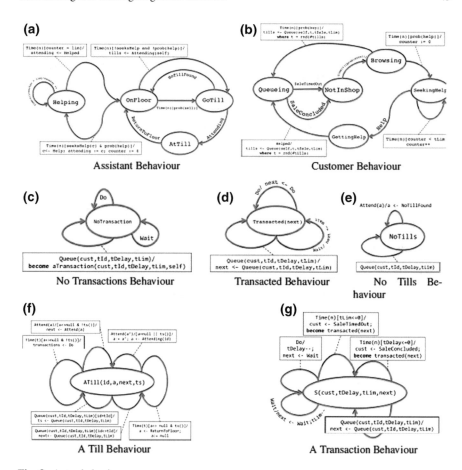

Fig. 9 Actor behaviours

Figure 11 shows a simple ESL query that interrogates the history produced by the shop simulation where `raid(n,cs)` is satisfied when `cs` is a list of n customer identifiers where the customers are all receiving help from sales assistants at the same time. Line 3 queries the history for all the customers and line 4 constructs a subset of all the customers. The rule *subset* is defined to allow backtracking through all the possible subsets, so line 5 will filter the subsets to select just those of the required length. The rules `findAllHelped` and `allHelped` query the history to ensure that the selected customers are in a `GettingHelp` state at the same time. The ESL implementation of the shop simulation can be downloaded as part of the open-source ESL system[1].

---

[1]https://github.com/TonyClark/ESL.

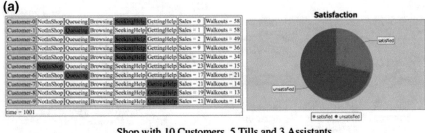

(a)

Shop with 10 Customers, 5 Tills and 3 Assistants

(b)

Shop with 10 Customers, 5 Tills and 5 Assistants

Fig. 10 Shop simulation output

```
1.   raid::(Int,[Int]);
2.   raid(n,raiders) <-
3.       customers(customers), !,
4.       subset[Int](raiders,customers),
5.       length[Int](raiders,n),
6.       findAllHelped(raiders);
7.
8.   findAllHelped::([Int]);
9.   findAllHelped(raiders) <- allHelped(raiders,t), !;
10.  findAllHelped(raiders) <- next[findAllHelped(raiders)];
11.
12.  allHelped::([Int],Int);
13.  allHelped([],t)      <- !;
14.  allHelped([c|cs],t) <-
15.      update(c,'state',GettingHelp,t),
16.      allHelped(cs,t);
```

Fig. 11 Raids: finding a pattern in a history

7   Conclusion

In this chapter we have sought to address the challenge of creating and analyzing actor-histories. We have shown how to extend a general actor-based language to produce histories of facts and how to extend a Prolog engine with temporal operators that can query the histories to establish whether patterns of facts exist. The proposal has been evaluated by showing that it can support a typical simulation, and that it can be implemented. Whilst other approaches to agent-based systems have used temporal operators to specify behaviour, and such operators have been used to interrogate system traces, use of this approach to analyze agent-based simulations is novel.

Whilst we have evaluated the approach in several different ways, the following issues and threats to validity remain: (Threat-1) The case study that has been used to evaluate the approach is taken from the literature and we can claim it to be representative of a class of simulations. Further work is needed to establish whether this case study is representative of a sufficiently broad class. (Threat-2) The histories that are produced by ESL-based simulations are typically intended to reflect some aspects of a real-world situation. Although the query approach described in this chapter does not rely on a valid history, in practice there must be some way to validate the simulation output. One possibility is to model an accepted theory, for example from social science or organisational management, and to show that the simulation model and its results is consistent with the theory. We intend to investigate this approach in the context of ESL simulations. (Threat-3) The efficiency of the approach has been established in the context of the example. This relies on knowledge of query semantics in order to ensure they are executed efficiently. It remains to be seen whether this is reasonable and whether efficiency can be improved. (Threat-4) Histories can be very large for long simulation runs. We have defined a compact implementation format, but further work is required to ensure that histories are no larger than is required. One option is to pre-process histories based on partial knowledge of query structures.

The approach described in this chapter is similar to existing approaches that analyze system traces, however we go further by defining precisely how the execution histories are produced and give a complete specification of the query language that can be used to analyze them. As such the work is novel and solves a problem that arises with actor-based systems where the behavior is both complex and non-deterministic. It remains to be seen whether the results meet users needs in terms of their ability to construct appropriate queries. One option is to be able to display and compare the results graphically, and this is an area for further work.

References

1. Ferber, J., Gutknecht, O.: A meta-model for the analysis and design of organizations in multiagent systems. In: Proceedings. International Conference on Multi Agent Systems, 1998., pp. 128–135. IEEE (1998)
2. Morgan, G.P., Carley, K.M.: An agent-based framework for active multi-level modeling of organizations. In: International Conference on Social Computing, Behavioral-Cultural Modeling and Prediction and Behavior Representation in Modeling and Simulation SBP-BRiMS 2016. Springer, Berlin (2016)
3. Pynadath, D.V., Tambe, M.: An automated teamwork infrastructure for heterogeneous software agents and humans. Auton. Agent. Multi-Agent Syst. 7(1–2), 71–100 (2003)

4. Fishwick, P.A.: Computer simulation: growth through extension. Trans. Soc. Comput. Simul. **14**(1), 13–24 (1997)
5. McDermott, T., Rouse, W., Goodman, S., Loper, M.: Multi-level modeling of complex sociotechnical systems. Proc. Comput. Sci. **16**, 1132–1141 (2013)
6. Clark, T., Kulkarni, V., Barat, S., Barn, B.: Sense-making in a complex and complicated world. IBM Syst. J. **42**(3), 462–483 (2003)
7. Barat, S., Kulkarni, V., Clark, T., Barn, B.: Enterprise modeling as a decision making aid: a systematic mapping study. In: The Practice of Enterprise Modeling—9th IFIP WG 8.1. Working Conference, PoEM 2016, Skövde, Sweden, pp. 289–298
8. Barat, S., Kulkarni, V., Clark, T., Barn, B.: A simulation-based aid for organisational decisionmaking. In: 11th International Joint Conference on Software Technologies (ICSOFT 2016)—Volume 2: ICSOFT-PT, Lisbon, Portugal, July 24–26, 2016, pp. 109–116
9. Barat, S., Kulkarni, V., Clark, T., Barn, B.: A model based realisation of actor model to conceptualise an aid for complex dynamic decision-making. In: Proceedings of the 5th International Conference on Model-Driven Engineering and Software Development, MODELSWARD 2017, Porto, Portugal, February 19–21, 2017, pp. 605–616
10. Clark, T., Kulkarni, V., Barat, S., Barn, B.: ESL: an actor-based platform for developing emergent behaviour organisation simulations. Advances in Practical Applications of Cyber-Physical Multi-Agent Systems: The PAAMS Collection—15th International Conference, PAAMS 2017, Porto, Portugal, June 21–23, 2017
11. Kulkarni, V., Barat, S., Clark, T., Barn, B.: A wide-spectrum approach to modelling and analysis of organisation for machine-assisted decision-making. In: Enterprise and Organizational Modeling and Simulation—11th International Workshop, EOMAS 2015, Held at CAiSE 2015, Stockholm, Sweden, June 8–9, 2015
12. Kulkarni, V., Barat, S., Clark, T., Barn, B.S.: Toward overcoming accidental complexity in organisational decision-making. In: 18th ACM/IEEE International Conference on Model Driven Engineering Languages and Systems, MoDELS 2015, Ottawa, ON, Canada, September 30–October 2, 2015, pp. 368–377 (2015)
13. Ricci, A., Agha, G., Bordini, R.H., Marron, A.: Special issue on programming based on actors, agents and decentralized control. Sci. Comput. Program. **98**, 117–119 (2015)
14. Clark, T., Kulkarni, V., Barat, S., Barn, B.: Actor monitors for adaptive behaviour. In: Proceedings of the 10th Innovations in Software Engineering Conference, ISEC 2017, Jaipur, India, February 5–7, 2017, pp. 85–95
15. De Koster, J., Van Cutsem, T., De Meuter, W.: 43 years of actors: a taxonomy of actor models and their key properties. In: Proceedings of the 6th International Workshop on Programming Based on Actors, Agents, and Decentralized Control, pp. 31–40. ACM (2016)
16. Hewitt, C.: Actor model of computation: scalable robust information systems. arXiv preprint arXiv:1008.1459 (2010)
17. Bosse, T., Jonker, C.M., Van der Meij, L., Sharpanskykh, A., Treur, J.: Specification and verification of dynamics in cognitive agent models. In: IAT, pp. 247–254. Citeseer (2006)
18. Caillou, P., Gaudou, B., Grignard, A., Truong, C.Q., Taillandier, P.: A simple-to-use BDI architecture for agent-based modeling and simulation. In: The Eleventh Conference of the European Social Simulation Association (ESSA 2015) (2015)
19. Galland, S., Knapen, L., Gaud, N., Janssens, D., Lamotte, O., Koukam, A., Wets, G., et al.: Multi-agent simulation of individual mobility behavior in carpooling. Transp. Res. Part C: Emerg. Technol. **45**, 83–98 (2014)
20. Singh, D., Padgham, L., Logan, B.: Integrating BDI agents with agent-based simulation platforms. Auton. Agent. Multi-Agent Syst. **30**(6), 1050–1071 (2016)
21. Bosse, T., Jonker, C.M., Van Der Meij, L., Treur, J.: LEADSTO: a language and environment for analysis of dynamics by simulation. In: German Conference on Multiagent System Technologies, pp. 165–178. Springer, Berlin (2005)

22. Bosse, T., Duell, R., Memon, Z.A., Treur, J., Van DerWal, C.N.: Multi-agent model for mutual absorption of emotions. ECMS **2009**, 212–218 (2009)
23. Sukthankar, G., Sycara, K.: Simultaneous team assignment and behavior recognition from spatio-temporal agent traces. AAAI **6**, 716–721 (2006)
24. Vasconcelos, W.W., Kollingbaum, M.J., Norman, T.J.: Normative conflict resolution in multiagent systems. Auton. Agents Multi-Agent Syst. **19**(2), 124–152 (2009)
25. Ndumu, D.T., Nwana, H.S., Lee, L.C., Collis, J.C.: Visualising and debugging distributed multi-agent systems. In: Proceedings of the IRD Annual Conference on Autonomous Agents, AGENTS '99, pages 326–333, New York, NY, USA, 1999. ACM
26. Bulling, N., Van der Hoek, W.: Preface: special issue on logical aspects of multi-agent systems. Stud. Log. (Special Issue), 2016 (2016)
27. Winikoff, M., Cranefield, S.: On the testability of BDI agent systems. J. Artif. Intell. Res. (JAIR) **51**, 71–131 (2014)
28. Borgwardt, S., Lippmann, M., Thost, V.: Temporal query answering in the description logic DL-lite. In: International Symposium on Frontiers of Combining Systems, pp. 165–180. Springer, Berlin (2013)
29. Al-Kateb, M., Ghazal, A., Crolotte, A., Bhashyam, R., Chimanchode, J., Pakala, S.P.: Temporal query processing in teradata. In: Proceedings of the 16th International Conference on Extending Database Technology, pp. 573–578. ACM (2013)
30. Kaufmann, M., Vagenas, P., Fischer, P.M., Kossmann, D., Färber, F.: Comprehensive and interactive temporal query processing with sap hana. Proc. VLDB Endow. **6**(12), 1210–1213 (2013)
31. Kaufmann, M., Manjili, A.A., Vagenas, P., Fischer, P.M., Kossmann, D., Färber, F., May, N.: Timeline index: a unified data structure for processing queries on temporal data in SAP HANA. In: Proceedings of the 2013 ACM SIGMOD International Conference on Management of Data, pp. 1173–1184. ACM (2013)
32. Kruse, R., Steinbrecher, M., Moewes, C.: Temporal pattern mining. In: 2010 International Conference on Signals and Electronic Systems (ICSES), pp. 3–8. IEEE (2010)
33. Räim, M., Di Ciccio, C., Maggi, F.M., Mecella, M., Mendling, J.: Log-based understanding of business processes through temporal logic query checking. In: OTM Conferences, pp. 75–92. Springer, Berlin (2014)
34. Borgwardt, S., Lippmann, M., Thost, V.: Temporalizing rewritable query languages over knowledge bases. In: Web Semantics: Science, Services and Agents on the World Wide Web, 50–70 (2015)
35. Koeman, V.J., Hindriks, K.V.: Designing a source-level debugger for cognitive agent programs. In: International Conference on Principles and Practice of Multi-Agent Systems, pp. 335–350. Springer, Berlin (2015)
36. Hindriks, K.V.: Debugging is explaining. In: International Conference on Principles and Practice of Multi-Agent Systems, pp. 31–45. Springer, Berlin (2012)
37. Agha, G.A.: Actors: a model of concurrent computation in distributed systems. Tech. rep., DTIC Document (1985)
38. Agha, G.A., Mason, I.A., Smith, S.F., Talcott, C.L.: A foundation for actor computation. J. Funct. Program. **7**(01), 1–72 (1997)
39. Karmani, R.K., Shali, A., Agha, G.: Actor frameworks for the JVM platform: a comparative analysis. In: Proceedings of the 7th International Conference on Principles and Practice of Programming in Java, pp. 11–20. ACM (2009)
40. Imam, S., Sarkar, V.: Savina-an actor benchmark suite. In: 4th International Workshop on Programming based on Actors, Agents, and Decentralized Control, AGERE (2014)
41. Gaintzarain, J., Lucio, P.: Logical foundations for more expressive declarative temporal logic programming languages. ACM Trans. Comput. Log. (TOCL) **14**(4), 28 (2013)
42. Siebers, P., Aickelin, U.: A first approach on modelling staff proactiveness in retail simulation models. J. Artif. Soc. Soc. Simul. **14**(2) (2011). URL: http://jasss.soc.surrey.ac.uk/14/2/2.html

Challenges in the Design of Decision Support Systems for Port and Maritime Supply Chains

Julio Mar-Ortiz, María D. Gracia and Norberto Castillo-García

Abstract The logistics in ports and maritime supply chains has reached a degree of complexity, that the management of supply chain operations requires of analytical methods to support with objectivity the decision-making process. For practical reasons, these analytical methods need to be embedded into technological platforms in the form of Decision Support Systems (DSS), in order to facilitate the required computations. Several DSS have been developed to address a variety of supply chain operational problems in the port and maritime industry. However, most of them set aside the fundamental discussion on which technological and analytical components are the most suitable for a particular problem. The purpose of this chapter is to survey the literature on the design and development of DSS for the port and maritime industry. We systematically review the works on existing methods (analytical and technological), and distinguish the gaps and tendencies of future developments in this industrial domain. We believe that the following DSS in the context of maritime transport will take advantage of the theoretical development of collaborative systems, data analytics and robustness to ease decision making process in the port and maritime industry. Implications to DSS developers for port and maritime supply chains are discussed.

Keywords Port operations management · Maritime supply chains
Decision support systems · Container terminals · Complex decision making
Multimodal logistics platforms

J. Mar-Ortiz (✉) · M. D. Gracia · N. Castillo-García
Faculty of Engineering, Universidad Autonoma de Tamaulipas,
Campus Tampico-Madero, 89140 Tampico, Mexico
e-mail: jmar@docentes.uat.edu.mx

© Springer International Publishing AG 2018
R. Valencia-García et al. (eds.), *Exploring Intelligent Decision Support Systems*,
Studies in Computational Intelligence 764,
https://doi.org/10.1007/978-3-319-74002-7_3

1 Introduction

In the recent years the logistics in port and maritime supply chains has reached such a degree of complexity that the management of port operations requires of analytical methods to support with objectivity the decision-making process. The use of advanced analytical methods during the long term or daily operations is difficult to implement without the aid of technological platforms that simplify the required computations. Therefore, Decision Support Systems (DSS) are required to aid decision makers in the complex decision-making process of managing port and maritime supply chains. DSS are computerized solutions used to assist the complex process of decision making and to solve specific problems [1]. In the port and maritime industry, a DSS can aid human decision making by providing access to relevant knowledge and supporting the choice among well-defined alternatives. DSS make use of analytical methods, artificial intelligence, operations research models and algorithms, simulation models, data mining models, modern statistical models and multi criteria decision theory methodologies to integrate robust technological platforms able to improve the decision making in complex systems.

From the early contributions of DSS (e.g. [2–4]) that use classical analytical models to solve structured and well-defined problems in container terminals; to the recent developments that use new technologies to get data from several sources like data files, internet, public data bases and sensors [5], and big data analytics techniques to convert data into useful information [6] in a collaborative environment that allows the interaction among several decision makers [7], through mobile cloud platforms to facilitate real-time communication [8]; the computational, analytical and technological advances have evolve to cope with the need of higher level intelligent and advanced systems that demand extensive experience and expert knowledge in the port and maritime industry.

Currently, the port and maritime industry requires robust DSS and models to facilitate information-sharing among several decision makers, to understand and analyze large volumes of reliable data, and to obtain qualified decisions on unstructured or semi-structured problems; where the environmental concerns and the need of dynamic models define the business environment. However, the existing methods, tools and techniques used to design and develop advanced DSS for the port and maritime industry remain largely unmapped nor discussed. Even though, a number of articles using a large variety of methods into a DSS to gather and study data related to decision-making in the port and maritime supply chains were published in the past decade, the main focus was frequently on the research results. However, most of them set aside the fundamental discussion on which technological and analytical components are the most suitable for a particular problem (this paper considers that a problem is defined by the data set under analysis and the decision to make). Accordingly, the purpose of this chapter is to survey the literature on the design and development of DSS for the port and maritime industry. Through a systematic approach, we identify the problems, the analytical methods used to analyze data, and the technological advances that outline the current paradigm for DSS in port and maritime supply chains. The discussion of these results allows us to distinguish the gaps and tendencies for future developments in the aforementioned industrial domain.

Decision support models have been developed for both public and private agents. Examples of private stakeholders are network operators, drayage operators, terminal operators and intermodal operators. Likewise, policy makers and port authorities are examples of public actors. Earlier review papers have studied several topics in port and maritime industry, where optimization techniques have been widely applied to solve vessel routing and scheduling problems, fleet management problems, bunkering problems, and the disruption management. Wang et al. [9] examines the literature on bunker optimization methods in maritime shipping. Caris et al. [10] reviews the development of DSS in intermodal transport, and classify the developments of DSS into six categories: policy support, terminal network design, intermodal service network design, intermodal routing, drayage operations, and information and communications technologies innovations. Christiansen et al. [11] reviews the developments of the vessel routing and scheduling problem. Mansouri et al. [12] reviews the role of multi-objective optimization as a DSS to improve the sustainability of maritime supply chains. Tran and Haasis [13] reviews optimization problems, methodologies and research directions concerning to network optimization in container shipping lines, with respect to container routing, fleet management and network design.

This survey is different from previous ones in aim, focus and scope. In this work, we concentrate on: (1) the methods used when looking for leveraging collective decision making—i.e. models to facilitate information-sharing among several decision makers and collaborative-data-driven DSS, (2) the methods used to obtain qualified decision on structured or semi-structured problems, (3) the methods used to understand and analyze large volumes of data—including methods for data gathering and analysis, and (4) the use of new technologies to enhance the collaborative decision making process. To the best of our knowledge, this is the first survey that focuses on these areas.

We summarize the main contributions of this chapter as follows:

- It identifies the problems, analytical methods used to analyze data, and technological advances that outline the current paradigm for DSS in port and maritime supply chains.
- It identifies the gaps in the design and development of DSS to enhance port and maritime supply chain operations.
- By means of a systematic search, our survey identifies the current literature in the following areas: (1) collective decision making, (2) problems and solution approaches, (3) data analytics methods, and (4) technological innovations for DSS.
- It provides a complete discussion of further research and developments for DSS in the port and maritime industry by methodically examining the literature in the previously mentioned areas and their overlap.

The remainder of this chapter is organized as follows. Section 2 analyzes the component of a DSS that settles down the bases of our discussion. The four research areas are reviewed in Sect. 3. Section 4 provides the discussion and implications for DSS developers. Finally, Sect. 5 summarize the main findings of the survey, specifying the main gaps, trends and future research directions.

2 Components of a Decision Support System

In the literature, there are three well-known methodologies to design and develop DSS (see Table 1). The system development life cycle (SDLC) methodology is a sequential procedure which starts by recognizing the goals of the system (needs of end users) and goes through different stages [14]. It is the most commonly used and hardest methodology. The rapid prototyping methodology [15] is a procedure that encourages faster development; it combines the efforts of the decision maker and the analyst. Both work closely to chart out specific requirements. The end-user DSS development methodology [16] promotes the design and development of a DSS based on the specific needs of a decision maker.

As expected, every approach has its advantages and disadvantages, where the analyst must consider all the corners of the problem. These drawbacks are generally the base to choose between do-it-yourself DSS versus professional development DSS. Professional developers are formally trained in computer science/business computing; on the other hand, do-it-yourself developer is a manager and end user

Table 1 Methodologies to design and develop DSS

System development life cycle approach	Rapid prototyping approach	End-user DSS development approach
• Identify the technical components required by the end user • Design the system architecture • Programming the DSS • Implement the DSS in the organization • Usage of the DSS • Evaluation and verification of the DSS functionality and capabilities • Performs the required adjustments or modifications	• Identify the end user requirements • Develop the first prototype • Evaluate the first prototype in order to identify the required adjustments and modifications • Test the DSS, and go back to evaluation, if needed • Implement the DSS	• Identify and bound the problem • Identify the users requirements • Integrate the team of experts according the expertise required • Identify the required solution approach • Design the software architecture • Identify the programming language • Prototype development Test and refinements of the prototype • Implementation Evaluation

who has an explicit knowledge of the problem but has typically little formal training in software and computer science.

The aim of a DSS is to aid decision makers in the decision-making process that includes three phases [17]: intelligence, design and choice. The intelligence phase consists of finding, identifying, and formulating the problem or situation that calls for a decision; the design phase involves the development of alternatives; finally, the choice phase contemplates evaluating the alternatives and selecting one for implementation. If all these three phases are well structured; i.e. there are well known procedures to formulate the problem, obtain solutions and evaluate the alternatives, we have a structured problem. But, if only some of these phases are well structured, requiring the individual judgment in some stages, then we have a semi-structured problem. On the other hand, if none of the phases are structured and the domain knowledge is mandatory in the decision-making process, then we have an unstructured problem [18].

Related to the design phase, there are two fundamental elements to consider when designing a DSS: (1) the components of the DSS, and (2) the type or category of the DSS. Regarding the components of a DSS there are two main references to consider: [19, 20]. Fanti et al. [19] identifies four components of a DSS): (1) the data component, (2) the model component, (3) the decision component, and (4) the interface component. The data component is responsible for the data gathering which may be internal or external. The model component includes the analytical models used to solve the problem. The decision component monitors the system state and, if required, runs a new scenario. The interface component is responsible for the communication and interaction of the system with the decision maker. Yazdani et al. [20] identifies three components of a DSS: (1) the data base management system, which serves as a data bank for the DSS; (2) the model-based management system, which is responsible to obtain solutions; and (3) the method of dialog generation and management system, which is responsible to interact with the user. The model-based management system can be divided into three phases: formulation, solution, and analysis. The formulation phase generates a model considering the assumptions and constrains of the current system. The solution phase solves the model by executing an algorithm. Finally, the analysis phase performs the 'what-if' analysis and interprets the solution or set of solutions.

Concerning the category of the DSS [14], identify six categories: (1) file drawer systems, (2) data analysis systems, (3) accounting and financial systems, (4) representational systems, (5) optimization systems, and (6) suggestion systems. A newer category scheme has been proposed by Power and Sharda [21]: (1) communications-driven systems, (2) data-driven systems, (3) document-driven systems, (4) knowledge-driven systems, and (5) model-driven systems.

Additionally, the analyst should consider that there are other design aspects that affect all DSS [22]: (1) the language system, (2) the presentation system, (3) the knowledge system, and (4) the problem processing system.

3 Methodology

We systematically review the existing literature to recognize the gaps in the design of DSS to aid decision making in port logistics chains and maritime supply chains, focusing on the methods, tools and techniques for designing, developing and implementing advanced DSS. As a basis for our research we use content analysis [23, 24], for which we define the following:

Material collection: in order to collect relevant publications regarding the design and implementation of DSS in port logistics chains and maritime supply chains, an extensive search of digital resources is performed in databases and publishers such as: ScienceDirect, Web of Science, Elsevier, Emerald, EBSCO, Inderscience, Springer, Hindawi, Taylor & Francis. The data bases were examined by using combinations of the keywords showed in Table 2. Thus, to find articles we combined (using AND) the keywords of the port logistics and maritime supply chains domain (using OR) with any of the keywords under the collaborative decision support systems domain (using OR). For instance: *port logistics chains* OR *container terminals* AND *decision support* OR *decision aid*. It should be noted that there were 441 combinations of the keywords in the scheme depicted above.

Aiming to enhance the consistency of the review process, each document was examinee and selected to be part of this chapter after fulfilling the following criteria: (1) exclusive reporting of DSS applications in port logistics and maritime supply chains; (2) exclusive use of DSS to support decision making of complex logistic and maritime operations, and (3) the paper was published in a scientific peer review journal.

Category selection and material evaluation: in order to define the formal aspects of the material to evaluate and to summarize the review results, the questions presented in Table 3 were formulated according to four categories or structural dimensions: (1) the journals and research groups, (2) problems and solution approaches, (3) data analytics methods, and (4) collective decision making and technological innovations. This allows identification of the issues or aspects and the interpretation of relevant results.

Table 2 Keywords used for searching data bases

Keywords related to port logistics and maritime supply chains	Keywords related to collaborative decision support systems
• Port logistics chains	• Decision support system
• Sea transport	• DSS
• Maritime logistics	• Decision support
• Port operations	• Decision aid
• Container terminals	• Decision tool
• Yard operations	• Decision making
• Vessel operations	• Collaborative DSS

Table 3 Category selection of structural dimensions

I. Journals and research groups	II. Problems and solution approaches
(1) How is the distribution of publications across the time period? (2) In which journals are such articles mainly published? (3) From which country are the research groups working in this topic? (4) What kind of paper is mainly published? (5) Which are the main aims of the DSS: monitoring, diagnostic, prescriptive, predictive?	(6) Which decision problems are addressed? (7) How the problem is formulated? (8) Which methods are used to deal with unstructured or semi-structured problems? (9) What solution approaches are applied? (10) What results and contributions were reached? (11) How the DSS aid the decision maker?
III. Data analytics methods	IV. Collective decision making and technological innovations
(12) Which data gathering methods and techniques are used? (How) (13) Characteristics of the what-if-analysis?	(14) Is the DSS collaborative? (15) Which methods are used when looking for leveraging collective decision making? (16) How the use of new technologies is incorporated into the DSS (social media, web semantic, linked data, big data, machine learning, IoT)?

4 DSS in Port and Maritime Supply Chains: State of the Art

4.1 Journals and Research Groups Descriptive Statistics

The material collection procedure initially produced more than 78 papers published in 30 scientific journals, this implies that 2.6 papers were published on average in each journal. A second-round selection results on a sample of 29 papers, published in 21 international scientific journals. The authorship of these papers was shared by 93 researchers. On the average there are 3.6 authors per paper, 14 (48.28%) of the papers were written by more than 3 authors, only two papers were written by a single author, whilst just one paper was authored by seven authors.

Figure 1 illustrates some important descriptive statistics from the reviewed papers. In order to classify the group of researchers working on the development of DSS for the port and maritime industry, we identify that in 17 papers the first author were from Europe, 7 from Asia, 3 from North America and 2 from South America. The first authorship of the reviewed papers was located in 15 different countries: 4 papers from England, 4 from Netherlands, 3 from the United States of America, 3 from Italy, and 3 from China; the first author of the remaining 12 papers was located in other 10 countries. All collected papers were published in different journals; more than 24% of papers were published on the Decision Support System journal. It was found that 58.6% of the papers were published on computer science journals, 27.6% were published on operational research journals, and 13.8% were

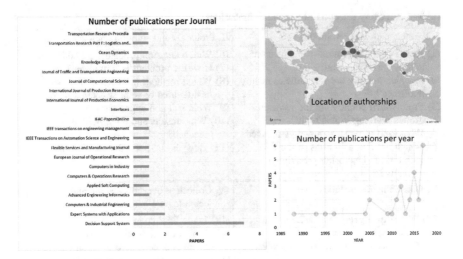

Fig. 1 Descriptive statistics of the reviewed papers

published on transportation science journal. It was found that 41.38% of the papers were published in the last two years.

All papers were classified into four port and maritime domains: container and bulk terminals (48%), shipping lines (20%), port and maritime management issues (16%) and interface port–container terminals (16%). It can be seen that classical DSS has been proposed to solve single structured problems in over 85% of the papers, whereas the remaining 15% have explored collaborative decision making, context-awareness decisions, and new technological issues. Regarding the main aims and objectives of the current DSS 76% are prescriptive only, 12% predictive and 4% monitoring, whilst the remaining 8% use a combination of them.

4.2 Problems and Solution Approaches

We focus our discussion on the port and maritime domains, aiming to identify the supply chain problems as well as the data-driven and analytical models used to support the decision-making process.

4.2.1 Shipping Lines Problems

DSS identified in the shipping lines domain are typically designed to aid vessel routing and scheduling problems. Relevant problems in this port and maritime domain are: the vessel routing and scheduling problem and the crew optimization problem. Most of these are structured problems, formulated using network optimization models, and solved via relaxation or population bio-inspired meta-heuristic algorithms like particle swarm optimization or genetic algorithms. The DSSs are mainly prescriptive.

The reviewed papers that fall into this category are: Ref. [3] which propose a DSS for optimizing the level of crew in vessels. The system identifies minimum safe manning levels for different types of ships and provides guidance and advice to the Coast Guard and ship owners in shaping minimum safe manning levels, given ship type, trade, technology level, etc. Data for the DSS was provided by American President Lines and Exon Shipping Company. The authors of the paper tested their DSS with real data from one containership from APL and one mixed-product tanker from Exxon Company. The authors do not specify any technical condition of their DSS.

Shen and Khoong [4] propose a DSS to solve a large-scale planning problem related to the multi-period distribution of empty containers for a shipping company. Their system uses network optimization models. Their system is able to determine cost-effective container leasing-in and off-leasing decisions. Kim and Lee [25] propose an optimization-based DSS for ship scheduling of bulk carriers. The ship scheduling problem is related to the movement of ships between a loading and a discharging port, then run empty until they reach the next loading port. Their prototype named MoDiSS (Model-based Support System in Ship Scheduling) is based on a set-packing model. The proposed DSS assigns bulk cargo to a schedule in a tramp shipping. The ship scheduling problem is formulated as a generalized set-packing problem and a Lagrangean relaxation algorithm is proposed to solve it. The optimization module of the DSS uses LINDO optimizer to solve the problem. Fagerholt [26] describes a commercial optimization-based DSS called TurboRouter. The author discusses the history of TurboRouter from its beginning as a research project to its commercialization. The programming language of TurboRouter is C++ version 6.0. The DSS presents interesting ways to show information and to interact with the user. TurboRouter is an optimization-based DSS for the vessel fleet routing and scheduling problem.

Recently, Lee et al. [6] propose a DSS for vessel speed optimization. Optimizing the speed of liner vessels significantly reduce fuel consumption and emissions. They propose a bi-criteria optimization model for a vessel routing problem. The optimization problem aims to minimize fuel consumption for vessels traveling through a predefined route while maximizing the total service level. A particle swarm optimization metaheuristic algorithm is proposed to solve the problem. The DSS is composed of the following components: the user interface, the weather archive data parser, the weather impact miner, and the solver. The user interface is a web-based system used to independently interact with end users. The weather archive data parser takes data from weather files and converts them into a data format that can be read by other components of the DSS. The weather impact miner aims at finding rules regarding weather impact. The solver uses the weather impact data to obtain Pareto optimal solutions and shows the trade-off relationships between the two criteria.

4.2.2 Container and Bulk Terminals

DSS identified in the container and bulk terminals domain are mostly designed to cranes assignment and scheduling problems. There is a vast list of relevant problems in this port and maritime domain, some of them are: the berth allocation problem, the dynamic container allocation problem, the yard and quay crane assignment and scheduling problems, and the internal trucks routing problems among others. Most of these are structured problems. Early stages in the development of DSS use mathematical programming and queuing models to address the problems. A second group of papers employ multi-criteria and robust optimization models to handle uncertainty. Recent developments use data analytic techniques based on artificial intelligence or data mining algorithms and expert systems, aiming to take real-time decisions in a context-aware environment. Therefore, metaheuristic algorithms may be used. The DSS are mainly prescriptive.

The reviewed papers that fall into this category are: Van Hee and Wijbrands [2] which describe a DSS for the capacity planning of container terminals. The paper devises models to determine the storage utilization, the yard stacker, berth utilization and queuing models. Their DSS can be used to support managers in strategic (e.g. quay capacity and yard layout) and tactical (e.g. the choice of the rules to allocate containers on the yard and to extend or reduce the number of trailers for the transport of containers from or to a ship) decisions.

Murty et al. [27] discuss the mathematical models and algorithms used in the design of a DSS aiming to solve nine of the daily operational problems in a container terminal: allocation of berths to arriving vessels, allocation of quay cranes to docked vessel, appointment times to truck carriers, routing of internal and external trucks within the terminal yard, dispatch policy at the terminal gate and the dock, storage space assignment, yard crane deployment, internal trucks allocation to quay cranes, and optimal internal trucks hiring plans. The DSS is inspired in some conditions faced in the terminals of Hong Kong. The architecture of this system can be found on Ref. [28].

Bandeira et al. [29] address the problem of the dynamic distribution and allocation of empty and full containers. The authors proposed two mathematical formulations and a DSS that runs exclusively on Microsoft Windows. The DSS uses several programming languages (not specified) and incorporates Lindo to solve the integer programming formulations. In addition, the DSS uses Microsoft Access for the database and Visual Basic for Applications (VBA) for the procedures. Liu et al. [30] describe three modules of a DSS, namely, demand forecasting, stowage planning, and shipping line optimization for a maritime container terminal. The system constructs problem models and uses exponential smoothing, regression analysis and neural networks for the demand forecasting issue; linear programming models for the stowage planning issue; and genetic algorithm and sequence alignment methods to solve the shipping line optimization problem. All problems are structured and the DSS may be classified as standalone for a single container terminal.

Ngai et al. [31] describe the development and application of an intelligent context-aware DSS for real-time monitoring of container terminal operations in a container terminal of Hong Kong. The intelligent context-aware DSS employs a ZigBee-based ubiquitous sensor with network technology. Through a case study they show the effectiveness of the system in supporting the real-time tracking and tracing of container trucks, quay cranes, and rubber-tired gantry cranes in a container terminal. Yan et al. [32] propose a DSS based on a knowledge-based system for a yard crane scheduling problem, based on the experience and knowledge of domain experts. In order to organize the acquired knowledge to make the rules extraction easier, the authors use a knowledge sorting process. Salido et al. [33] propose a DSS to aid terminal operators in the transshipment (loading/unloading tasks) of containers between vessels and land vehicles. The efficiency of this operation depends on the number of assigned cranes and the efficiency of the container yard logistic.

Ursavas [34] proposes a DSS to optimize operations on the quay side of a container terminal. A bi-objective optimization model for the combined berth allocation, crane assignment and scheduling problems is proposed. The software is designed to run in a PC environment under Microsoft Windows. The main components of the DSS contain the user interface for input parameters, user re-evaluation and report visualization, the data base management system for storing the data related with the terminal such as vessels, berth structures and cranes and the core of the DSS where the model is solved in compliance with the solution algorithm. Microsoft Access is used to manage the database. Gurobi solver is used for solving the procedures and the interaction with the optimization algorithm is developed in Visual Basic. With regard to the requirements of the decision makers, the flexibility of the DSS has been seen as an important value.

Kapetanis et al. [35] describe a DSS fed with input from tracking and tracing systems or traffic monitoring systems, to perform a transportation plan for the Piraeus container terminal. van Riesen et al. [36] derive real-time decision rules for suitable allocations of containers to inland services. They propose a general four step method: historic data assembly, optimization of historic sets, decision tree inference, and application of the decision tree in real time. The proposed method uses an offline optimal planning method to get optimal results of historic transportation problems. The results are translated into a decision tree with the inference method. Also, a mathematical model for obtaining multiple offline optima with equivalent objective value as input for the learning algorithm is used.

Recently, de León et al. [37] use an artificial intelligence technique to solve a scheduling problem in the context of the maritime terminal operations. The DSS uses a machine learning approach to select an algorithm that best suits the given instance. The DSS was trained by running a set of algorithms on a set of training instances. Thus, the aim of this DSS is to improve the recommendation of algorithms for a given instance by the meta-learning technique. This technique takes into account data from past instances with similar features and the knowledge of the performance of the available algorithms on these instances.

Heilig et al. [8] describe a mobile cloud platform for the real-time optimization of transport of containers between transshipment areas. The authors argue that both real-time information exchange and optimization are necessary to efficiently coordinate actors and container movements in inter-terminal transport. They formulate the inter-terminal truck routing problem as an optimization problem and propose hybrid simulated annealing algorithms to solve it. The optimization component is embedded into a cloud platform that integrates both real-time data gathering from truck drivers using a mobile app and current traffic data. Their proposed mobile cloud platform acts as a DSS facilitating real-time communication and context-aware planning for inter-terminal transportation.

4.2.3   Interface Port–Container (Bulk) Terminals

DSS identified in the interface port–container or bulk terminals domain are mostly designed to manage the flows between the inland port and the seaport. Most of these are semi-structured problems that use the fuzzy set theory or any type of multi-criteria decision analysis approach to synthesize information. The DSSs are mainly prescriptive and exploit the collaborative decision-making paradigm.

The reviewed papers that fall into this category are: Refs. [38, 39] which develop a DSS for robust berth allocation in bulk material handling ports using a genetic algorithm. The problem is formulated as bi-objective berth allocation problem to minimize the operational delay and customer priority deviation. The DSS is developed for the port authority.

Fanti et al. [19] present a simulation-optimization based DSS to manage the flow of goods and business transactions between the inland port and the seaport. Fazi et al. [40] propose a DSS for the hinterland allocation problem. The authors mathematically modeled the problem and solved this model with the metropolis algorithm, which is a variant of the well-known simulated annealing algorithm. The authors used a case study to validate the model and the DSS. They suggested that a heuristic with a few parameters is recommended since decision makers require easy parameterization. That is the main reason why they selected the metropolis algorithm to be included in the DSS. This algorithm has only one parameter to be tuned: the temperature.

4.2.4   Port and Maritime Management Issues

DSS identified in the port and maritime management issues domain are mostly designed to evaluate risk factors within the ports, and selecting appropriate resilience strategies. Most of these are semi-structured problems that use the fuzzy logic and multi-criteria decision analysis approaches, or a combination of both like the fuzzy analytical hierarchical process. The DSSs are mainly predictive and exploit the collaborative decision-making paradigm.

The reviewed papers that fall into this category are: Mokhtari et al. [41] which propose a decision support framework for risk management on sea ports and terminals. They use the fuzzy set theory to describe and evaluate the associated risk factors within the ports and terminals operations and management; and the evidential reasoning approach to synthesize the information produced. Three cases of study in Iranian ports of Bushehr, ShahidRajaie and Chabahar were used to prove the applicability of the proposed approach. This is not a DSS, it is just a decision support framework.

Grasso et al. [42] propose a DSS for evaluating the risk concerning the weather conditions for civilian and military maritime activities. The DSS uses a fuzzy logic classifier to handle the uncertainty in the inputs of the weather conditions. John et al. [43] propose a collaborative DSS based on a fuzzy analytical hierarchical process, which is used to analyze the order preference in a complex structure in a group decision making process, in order to select an appropriate resilience strategy for a seaport system's operation. Guarnaschelli et al. [44] propose a DSS tool for managing the export process in the wood product enterprise. The problem approached in this paper has several objectives, which sets a multi-criteria optimization problem. The authors used the lexicographic goal programming approach to solve the multi-criteria optimization problem. They validated the DSS by means of a case study.

Recently, Fanti et al. [7] describe a cloud-based cooperative DSS for intermodal transportation systems and terminal operators, and propose a general framework to develop collaborative DSS. Their framework involves two steps. The user requirement collection step, that embraces a survey process aiming at identifying stakeholders' expectations. And the design step, that transforms the users' requirements into systems requirements.

From the preceding analysis, it can be observed that most of the DSS published in the literature address structured problems. Therein, mathematical programming and analytical models are commonly used to support the decision-making process. In the shipping lines domain, vessel routing and scheduling problems take advantage of network optimization and generalized set-packing models. In the container and bulk terminals domain, queuing models are considered to address congestion issues. In both domains, it seems a growing trend the use of multi-criteria and robust optimization models to handle uncertainty. In the third and four domains, the interface port–container terminals and the port and maritime management issues, respectively, semi-structured problems arise. These use the fuzzy set theory or any type of multi-criteria decision analysis approach to synthesize information. However, to the best of our knowledge, none DSS have been proposed to address unstructured problems in the port and maritime industry. However, as stated by Widz et al. [45] when dealing with semi-structured problems, a DSS may combine data mining algorithms and expert systems within a simple interface to allow domain experts to modify the model components based on their experience. On the other hand, unstructured problems require extensive efforts on data gathering, preparation and visualization issues. In this case, the aim is to provide to the decision maker with enough information that combined with their domain knowledge will aid to get a solution.

4.3 Data Analytics Methods

The continuous growing of information and knowledge that should be mastered by decision makers in the port and maritime industry emphasizes the need for data-driven DSS using advanced and modern technologies. The big data predictive analytics is a term used to refer to a set of techniques destined to handle big data, characterized in terms of high volume, velocity and variety [46]. Big data can help to address critical challenges of predictive analytics that refer to data capture, storage, transfer and sharing (i.e. system architecture), and search, analysis, and visualization (i.e. data analytics). Hazen et al. [47] review the use the big data and predictive analytics tools and methodologies to improve operational and strategic capabilities of supply chains. As a relatively new and innovative business phenomenon, firms seek to leverage decision-making outputs from big data predictive analytics in their supply chains with the initial and highly visible goal of enhancing their financial bottom line.

The use of big data techniques embedded into a DSS may support the decision-making process in the port and maritime industry by analyzing data available on the internet through cognitive maps [5], or by employing random forecast regression methods to predict the profitability of insurance in containers [48]. Psaraftis et al. [49] survey the existing literature on dynamic vehicle routing problems. They recognize the importance of big data in vehicle routing problems to improve decision making. The new technologies applied on a collaborative scheme will produce large and more complex amount of data that need to be processed using big data analytics.

A survey of official online sources of high-quality free-of-charge geospatial datasets to support the design of maritime geographic information systems applications aiming to aid the maritime industry and its related organizations, is presented in [50]. The available datasets were empirically tested and their quality and usefulness verified, producing a selective thesaurus. The authors discuss the various classes of available data.

From the previous analysis, it can be stated that most of the DSS published on the literature use historical records from a data base. This could be explained because most of the papers dealt with structured problems, which employ well-defined procedures on well-defined data sets. Only a few of the most recent contributions gather data available on the internet. But, none of the papers describes the use of sensors to gather data from the environment. Note that the what-if analysis is a key functionality of any DSS and its use can be classified into three categories: (1) using what-if analysis to measure the impact of data uncertainties on performance indicators; (2) using what-if analysis for comparing alternative solutions; and (3) using what-if analysis to analyze trade-off relationships of conflicting performance indicators.

4.4 Collective Decision Making and Technological Innovations

In the middle of the last decade, Giannopoulos [51] argues that the most relevant technologies for the management of maritime traffic were: global positioning system (GPS), a satellite based navigation system which provides continuous position; electronic chart display and information systems (ECDIS), a system to operate and visualize digital navigation chart data; satellite-based communication systems with almost global coverage; geographical information systems (GIS); intelligent transportation systems (ITS); and automated identification systems (AIS), used for the automatic identification of a vessel and its coordinates. Currently, the technological resources to support the management of ports include a number of applications such as: optimization of support services such as pilot, tug and mooring services; improved organization of services such as custom services; optimization of port and terminal services such as berth, loading and discharge services; organization of shipping company and agent services among others.

According to [49] there are four critical elements that DSS will embrace in the forthcoming years: (1) advances in information and communication technologies (like social media, semantic web, linked data, big data, machine learning, web service and cloud computing) and related technologies (e.g. RFID, sensors); (2) advances in computer power and portability; (3) big data and predictive analytics methods; and (4) parallel and GPU programming. All these elements will facilitate the computation of optimal decisions a million times faster than 20 years ago, which is critical for specific class of problems in the port and maritime industry; mainly because their dynamic nature would demand such technologies on the one hand, and make optimal use of these technologies on the other.

RFID has been deployed by supply networks to improve asset utilization, effectively combat counterfeiting, and advance targeted product recalls. However, new affordable and deployable technologies and micro-sensors have recently appeared and keep maturing. Musa et al. [52] discuss the roles of the cyber-infrastructure, the real-time information flow and coordination mechanism to efficiently manage the dynamic of smart enterprises; as well as the need in logistics for embedded devices that integrates RFID and sensors for positioning and environmental parameters; the support technologies and services that might be implemented in such a device. They outline the development of an embedded micro system that combines RFID, GPRS, GPS and environmental sensors for applications in logistics.

Effective and instantaneous information flow in a supply chain network is now feasible because of the ever-increasing growth, speed, convergence and adoption of the internet, wired and wireless infrastructure-based communication networks, sensor networks, RFID, ubiquitous positioning, micro electromechanical systems, software-defined radios, chip-scale atomic clocks, mobile computing, semantic web services, intelligent software agents, especially mobile multi-agents. Therefore, the use and development of modern technologies and collaborative decision-making are closely interrelated. The need to design collaborative DSS was recognized by

Shen et al. [4], which argue that the globalization of shipping businesses will require group decision support systems and other computer-supported cooperative work software. Currently, communications technologies are centered in communications-driven DSS for supporting decision-making; however, there are vast and challenging opportunities to develop and integrate advanced computer technologies in the port and maritime industry.

Currently, the dynamics of the decision-making process in several industry domains is moving from individual decision makers to collaborative group decision makers; where the knowledge of several geographically distributed workers is considered as an asset of the organization. Shim et al. [1] identify three trends on DSS design. Firstly, several key technological advances have arose in the area of DSS. On one side there are the on-line analytical processing and data mining techniques. Some of the new technologies that will remain leading the major advances in DSS are mobile tools, mobile e-services, Web-based DSS and wireless Internet protocols. Secondly, the design of collaborative support systems and virtual teams are evolving from DSS to support individual decision-making, new DSS technologies are applied to workgroups or teams, especially virtual teams. Another trend is the application of state-of-the-art optimization models and algorithms. Finally, group supports systems or collaboration support systems improve the communication-related actions of group members concerned in cooperative work.

In the recent years, several studies focus on the design of DSS applications for the port logistics and maritime supply chains. Most of these studies can be considered as standalone company developments, where decisions concerns to the task of only one actor in the maritime industry (e.g. shipping lines, container terminals, port authorities, or hinterland actors). However, the introduction of new information and communication technologies has led to a new kind of DSS, which are called collaborative DSS. A collaborative DSS uses a cooperative approach among different stakeholders pursuing share objectives, in order to share information and data to get decisions. According to [7] collaboration occurs within the framework of cooperative work and is defined as "multiple individuals working together in a planned way in the same or in different but connected production processes". While coordination involves actors working together harmoniously to accomplish a collective set of tasks.

Collaboration has been one of the main aims in the port and maritime industry. Lei et al. [53] mathematically model and empirically investigate the operational performance of container-vessel schedules under three management policies: the non-collaborative policy, the slot-sharing policy, and the total-sharing (the total collaboration) policy. The study reveals that without partner carriers' full commitment to share the demand and the resources in a flexible manner, the advantage of collaborative planning cannot be fully exploited.

Ascencio et al. [54] propose a collaborative logistics framework for a port logistics chain based on the principles of supply chain management that rely on stakeholders' integration and collaboration. Their collaborative framework comprises three main dimensions: (1) the management of port logistic governance, (2) the logistics management platform system and (3) the port logistics operations model. The framework is focused on the design and implementation of the three inter-enterprise business processes of the port logistics chain: demand management, orders management, and vehicles management.

Feng et al. [55] describe the development of an agent-based information integrated platform for hinterland waterway barge transport planning. The objective is to facilitate data exchange and establish an optimal plan for involved parties. The agents interact via a web interface with the hinterland transportation planning system, where the agent behavior is coordinated, the level of information exchange is controlled and the plan is created and stored.

From the previous analysis, it can be stated that although [49] listed some of new technologies that DSS will need to embrace in the forthcoming years (like social media, semantic web, linked data, big data, machine learning, web service, cloud computing and IoT), to the best of our knowledge none of these new technologies have been incorporated into published DSS. This could be explained due to the fact that most of the papers deal with structured problems that requires specific input data that are encapsulated into well-defined data sets. It is expected that, as DSS begin to deal with unstructured problems, the data gathering phase will make use of these new technologies to collect reliable information. But at current time this is in an early stage.

However, the technological and analytical components that seems to be the most suitable for port and maritime supply chains are: social media, linked data, big data, machine learning, web service, cloud computing and IoT. The social media technology will be required to create and share information between terminal operators, shippers and trucking companies; this will particularly necessary as port community systems formalize. The linked data technology will be required to share information in a structured way to be read automatically by interconnected DSS; this will be particularly necessary to design complex and integrated DSS, what is a step ahead the current DSS that are presented as standalone solutions to particular problems. Evidently, big data is a current need to deal with the large amount of data daily captured in the maritime industry. The web service technology, cloud computing and IoT are currently used in some DSS reported in the literature [7, 8] to connect electronic devices for real time monitoring and optimization.

5 Discussion and Implications

5.1 Discussion

Ports act as a crossing point between ships and seashore by providing protection and a berthing space, momentary storage, and infrastructure for cargo handling operations within the port. Current trends in the evolution of global transport chains urges stakeholders in the logistics networks to re-think their roles within port communities [56]. Consequently, ports need to go beyond their traditional transshipment role and become more integrative.

Current challenges on the design of DSS to support supply chain operations in the port and maritime industry, are related to the design of collaborative DSS and the integration of modern information and communication technologies. This is a consequence of: (1) the need to improve the collaboration among stakeholders in a port community system in order to increase response effectiveness; (2) the use of information and communication technological solutions to support the coordination activities and the distributed decision-making processes; and (3) the need to handle large volumes of data from the operational records, the official data base available on the internet, and the sensor allocated along the supply network. Accordingly, there are five main elements that any further DSS development in the port and maritime industry should include, which define the gaps and directions for future research:

- A variety of artificial intelligence techniques and operational research methods to solve the decision problems, as well as a meta-learning technique (e.g. machine learning models) to select an algorithm that best suits the given instance [37]. Consider that algorithms with few parameters are recommended since decision makers require easy parameterization in DSS [40].
- Multi-criteria optimization models to consider the structural dimensions of decision makers in a collective decision-making process, as well as a complete vision of sustainability along the decision process [12].
- An efficient user interface based on web systems for efficient and independent interaction with end users [6]. Fagerholt [26] states that although most of the research efforts in vessel fleet scheduling have been focused on the development of optimization algorithms, his experience reveals that the ocean shipping industry was not yet ready for this. Therefore, he suggests to focus on the user interaction when developing a DSS. It should be noted that the flexibility of a DSS has been seen as an important value for the requirements of the decision makers [34]. Decision makers in a multi-enterprise collaborative environment need flexible systems that allow for seamless integration among all members of organizational supply networks without being dependent on the knowledge of the users.
- Efficient mechanisms to obtain data and transform them into a data format that can be read by other components of a DSS. Fagerholt [26] discusses the importance of integrating DSS with existing shipping company's administrative systems.
- The use of cloud and mobile cloud platforms to support cooperative DSS [7] and the real-time information exchange and optimization [8] which are necessary to efficiently coordinate actors and container movements along the maritime supply chains. A DSS must facilitate real-time communication and context-aware decision making. Finally, there is a trend in the design of ubiquitous computing systems, that allow virtual teammates truly collaborate anywhere and anytime, without the need to physically beat a computer tied to a wired network.

5.2 Implications

Identifying the problem and bound the reach of the decision making is the first step in the design and implementation of a DSS. A diagnostic study of the decision-making process may help to gain clarity on: (1) which is the main problem, (2) what goals are associated with the decision problem, (3) how to collect and which collection techniques to use, (4) how the decisions are made, and (5) who are involved in the decision making process. The interested reader should not overlook the fact that DSS can address more than one generic application, they usually involve multiple problems within the port and maritime industry that are interrelated and are interdependent.

Another important element in the design of a DSS is to know the expectations of the end user or the decision-maker who will use the DSS. Usually, a DSS is required for (1) monitoring, (2) diagnostic, (3) descriptive analysis, (4) prescriptive analysis, or (5) predictive analysis. Additionally, there are other dimensions to consider as the availability of resources and infrastructure, the quality of the solution accepted, or the need to obtain collective decisions. This will define the kind of DSS needed: communication-driven DSS, data-driven DSS, document-driven DSS, knowledge-driven DSS, model-driven DSS, or a combination of them.

With this information at hand, the qualifications required by the team group will be defined. For example, if the application requires mobile device connectivity, then an app developer will be required within the development team. If a model-driven DSS is required, then experts on the use of optimization, simulation or statistical methods will be required. On the other hand, if a data-driven DSS is needed to emphasize the access to and manipulation of large amounts of structured data, a big data analyst will be required. Note that multi-criteria decision analysis tools, like the analytical hierarchical process, may be used to build collaborative DSS, providing a framework for group decision making in a flexible manner within a complex and semi-structured problem domain.

The reader should consider that collaborative DSSs require a centralized coordination platform system and a collaborative knowledge base system, which is a communication module composing of several elements, representing various functions that the module should perform for the problem-solving stages. Figure 2 identifies the core components of a DSS.

The data component is responsible for the collection data to feed the model component. Usually, the historical data represent structural data, these data can be collected from external databases such as: truck carriers, highway authorities, shipping agents, terminal managers and port authorities. The real-time data come from smart devices and sensors that monitor environments, traffic, and weather conditions. A critical factor for the success of a DSS is to ensure that the information infrastructure in the terminal can generate all the data needed by the algorithms implemented in the DSS [28]. The model component is in charge to suggest and support the decision making during the decision process. It can merge

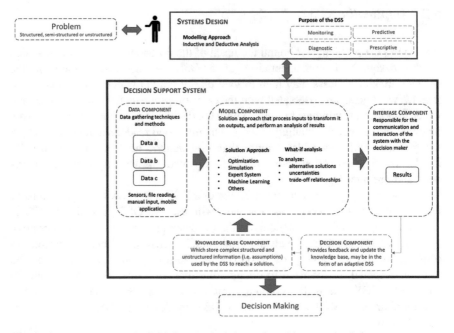

Fig. 2 Core components of a DSS for port logistics and maritime supply chains

information coming from the data and knowledge base components, in order to propose solutions to the decision maker through the interface component. The basic requirements of the interface component are: (1) efficiency, i.e. reliable, rapidly accessible information; a single access point; and the potential to personalize information requests, and (2) simplicity, i.e. clear content; easily understandable interfaces; and ease of use. The knowledge-based component includes four phases, i.e., the knowledge acquisition, knowledge sorting, rules extraction and knowledge reasoning.

6 Conclusions

This chapter reviewed the literature on the design and development of Decision Support Systems for the port and maritime industry, regarding supply chains operations. We found that the investigation on DSS for the port and maritime industry is growing rapidly. There is a need to develop portable, collaborative, context-awareness and robust DSS, able to handle and analyze on a real-time basis the large amount of data generated in the decision-making process in real time. Therefore, several research avenues are depicted. The first one is related to the design of portable DSS, which can be approached via cloud computing and ubiquitous computing. The design of collaborative DSS may be approached via multi-criteria decision-making models. This requires the development of computer technology to allow connectivity via web-services and cloud computing. The design of context-awareness DSS requires the development of sensors and access to internet public data bases. The design of robust DSS is required to handle the dynamics and uncertainties involved in the port and transport times. Moreover, the use of cutting-edge big data techniques allows to handle large amount of data. Finally, the design of DSS that provide real-time solutions may be approached using modern metaheuristic algorithms whose effectiveness has been demonstrated in several domains.

Acknowledgements This research was supported by the PRODEP program of the Mexican Minister of Education (SEP), through the Research Grants DSA/103.5/15/14164 (Research Network on Supply Chain Modeling and Optimization) and DSA/103.5/16/15436 (Postdoctoral Research Fund). The author of this chapter acknowledges the valuable comments of the referees to improve the quality of our work.

References

1. Shim, J.P., Warkentin, M., Courtney, J.F., Power, D.J., Sharda, R., Carlsson, C.: Past, present, and future of decision support technology. http://www.sciencedirect.com/science/article/pii/S0167923601001397 (2002)
2. van Hee, K.M., Wijbrands, R.J.: Decision support system for container terminal planning. Eur. J. Oper. Res. **34**, 262–272 (1988)
3. Grabowski, M., Hendrick, H.: How low can we go?: Validation and verification of a decision support system for safe shipboard manning. IEEE Trans. Eng. Manag. **40**, 41–53 (1993)
4. Shen, W.S., Khoong, C.M.: A DSS for empty container distribution planning. Decis. Support Syst. **15**, 75–82 (1995)
5. Choi, Y., Lee, H., Irani, Z.: Big data-driven fuzzy cognitive map for prioritising IT service procurement in the public sector. Ann. Oper. Res. 1–30 (2016)
6. Lee, H., Aydin, N., Choi, Y., Lekhavat, S., Irani, Z.: A decision support system for vessel speed decision in maritime logistics using weather archive big data. http://www.sciencedirect.com/science/article/pii/S0305054817301429 (2016)
7. Fanti, M.P., Iacobellis, G., Nolich, M., Rusich, A., Ukovich, W.: A decision support system for cooperative logistics. IEEE Trans. Autom. Sci. Eng. **14**, 732–744 (2017)
8. Heilig, L., Lalla-Ruiz, E., Voß, S.: Port-IO: an integrative mobile cloud platform for real-time inter-terminal truck routing optimization. Flex. Serv. Manuf. J. **29**, 504–534 (2017)
9. Wang, S., Meng, Q., Liu, Z.: Bunker consumption optimization methods in shipping: a critical review and extensions. Transp. Res. Part E Logist. Transp. Rev. **53**, 49–62 (2013)
10. Caris, A., Macharis, C., Janssens, G.K.: Decision support in intermodal transport: a new research agenda. Comput. Ind. **64**, 105–112 (2013)
11. Christiansen, M., Fagerholt, K., Nygreen, B., Ronen, D.: Ship routing and scheduling in the new millennium. http://www.sciencedirect.com/science/article/pii/S0377221712009125 (2013)
12. Mansouri, S.A., Lee, H., Aluko, O.: Multi-objective decision support to enhance environmental sustainability in maritime shipping: A review and future directions. Transp. Res. Part E Logist. Transp. Rev. **78**, 3–18 (2015)
13. Tran, N.K., Haasis, H.-D.: Literature survey of network optimization in container liner shipping. Flex. Serv. Manuf. J. **27**, 139–179 (2015)

14. Stodolsky, D., Alter, S.L.: Decision support systems: current practice and continuing challenges. Behav. Sci. **27**, 91–92 (1982). (Reading, Massachusetts: Addison-Wesley Publishing Co., 1980, 316 pp)
15. Tripp, S.D., Bichelmeyer, B.: Rapid prototyping: an alternative instructional design strategy. Educ. Technol. Res. Dev. **38**, 31–44 (1990)
16. Turban, E., Aronson, J.E., Liang, T.-P.: Decision support systems and intelligent systems. Pearson/Prentice Hall (2005)
17. Simon, H.A., Dantzig, G.B., Hogarth, R., Plott, C.R., Raiffa, H., Schelling, T.C., Shepsle, K. A., Thaler, R., Tversky, A., Winter, S.: Decision making and problem solving. Interfaces (Providence) **17**, 11–31 (1987)
18. Averweg, U.R.: Decision Support Systems and Decision-Making Processes. In: Encyclopedia of Decision Making and Decision Support Technologies, pp. 218–224. IGI Global (1), AD
19. Fanti, M.P., Iacobellis, G., Ukovich, W., Boschian, V., Georgoulas, G., Stylios, C.: A simulation based decision support system for logistics management. J. Comput. Sci. **10**, 86–96 (2015)
20. Yazdani, M., Zarate, P., Coulibaly, A., Zavadskas, E.K.: A group decision making support system in logistics and supply chain management. Expert Syst. Appl. **88**, 376–392 (2017)
21. Power, D.J., Sharda, R.: Decision support systems. In Springer Handbook of Automation. pp. 1539–1548. Springer Berlin Heidelberg, Berlin, Heidelberg (2009)
22. Bonczek, R.H., Holsapple, C.W., Whinston, A.B., Schmidt, J.W.: Foundations of Decision Support Systems. Elsevier Science (2014)
23. Brewerton, P., Millward, L.: Organizational Research Methods : A Guide for Students and Researchers. SAGE (2001)
24. Seuring, S., Müller, M.: From a literature review to a conceptual framework for sustainable supply chain management. J. Clean. Prod. **16**, 1699–1710 (2008)
25. Kim, S-H., Lee, K-K.: An optimization-based decision support system for ship scheduling. Comput. Ind. Eng. **33**, 689–692 (1997)
26. Fagerholt, K.: A computer-based decision support system for vessel fleet scheduling—experience and future research. Decis. Support Syst. **37**, 35–47 (2004)
27. Murty, K.G., Liu, J., Wan, Y.W., Linn, R.: A decision support system for operations in a container terminal. Decis. Support Syst. **39**, 309–332 (2005)
28. Murty, K.G., Wan, Y., Liu, J., Tseng, M.M., Leung, E., Lai, K.-K., Chiu, H.W.C.: Hongkong international terminals gains elastic capacity using a data-intensive decision-support system. Interfaces (Providence) **35**, 61–75 (2005)
29. Bandeira, D.L., Becker, J.L., Borenstein, D.: A DSS for integrated distribution of empty and full containers. Decis. Support Syst. **47**, 383–397 (2009)
30. Liu, Y., Zhou, C., Guo, D., Wang, K., Pang, W., Zhai, Y.: A decision support system using soft computing for modern international container transportation services. Appl. Soft Comput. J. **10**, 1087–1095 (2010)
31. Ngai, E.W.T., Li, C.-L., Cheng, T.C.E., Lun, Y.H.V., Lai, K.-H., Cao, J., Lee, M.C.M.: Design and development of an intelligent context-aware decision support system for real-time monitoring of container terminal operations. Int. J. Prod. Res. **49**, 3501–3526 (2011)
32. Yan, W., Huang, Y., Chang, D., He, J.: An investigation into knowledge-based yard crane scheduling for container terminals. Adv. Eng. Informatics. **25**, 462–471 (2011)
33. Salido, M.A., Rodriguez-Molins, M., Barber, F.: A decision support system for managing combinatorial problems in container terminals. In Knowledge-Based Systems, pp. 63–74. Elsevier (2012)
34. Ursavas, E.: A decision support system for quayside operations in a container terminal. Decis. Support Syst. **59**, 312–324 (2014)
35. Kapetanis, G.N., Psaraftis, H.N., Spyrou, D.: A Simple synchro—modal decision support tool for the piraeus container terminal. In Transportation Research Procedia, pp. 2860–2869. Elsevier (2016)
36. van Riessen, B., Negenborn, R.R., Dekker, R.: Real-time container transport planning with decision trees based on offline obtained optimal solutions. Decis. Support Syst. **89**, 1–16 (2016)

37. de León, A.D., Lalla-Ruiz, E., Melián-Batista, B., Marcos Moreno-Vega, J.: A machine learning-based system for berth scheduling at bulk terminals. Expert Syst. Appl. **87**, 170–182 (2017)
38. Pratap, S., Nayak, A., Cheikhrouhou, N., Tiwari, M.K.: Decision support system for discrete robust berth allocation. In IFAC-PapersOnLine, pp. 875–880. Elsevier (2015)
39. Pratap, S., Nayak, A., Kumar, A., Cheikhrouhou, N., Tiwari, M.K.: An integrated decision support system for berth and ship unloader allocation in bulk material handling port. Comput. Ind. Eng. **106**, 386–399 (2017)
40. Fazi, S., Fransoo, J.C., Van Woensel, T.: A decision support system tool for the transportation by barge of import containers: a case study. Decis. Support Syst. **79**, 33–45 (2015)
41. Mokhtari, K., Ren, J., Roberts, C., Wang, J.: Decision support framework for risk management on sea ports and terminals using fuzzy set theory and evidential reasoning approach. Expert Syst. Appl. **39**, 5087–5103 (2012)
42. Grasso, R., Cococcioni, M., Mourre, B., Chiggiato, J., Rixen, M.: A maritime decision support system to assess risk in the presence of environmental uncertainties: the REP10 experiment. Ocean Dyn. **62**, 469–493 (2012)
43. John, A., Yang, Z., Riahi, R., Wang, J.: Application of a collaborative modelling and strategic fuzzy decision support system for selecting appropriate resilience strategies for seaport operations. J. Traffic Transp. Eng. (English Ed. 1), 159–179 (2014)
44. Guarnaschelli, A., Bearzotti, L., Montt, C.: An approach to export process management in a wood product enterprise. Int. J. Prod. Econ. **190**, 88–95 (2017)
45. Widz, S., Ślęzak, D.: Rough set based decision support—models easy to interpret. In Rough Sets: Selected Methods and Applications in Management and Engineering, pp. 95–112. Springer, London (2012)
46. Gunasekaran, A., Papadopoulos, T., Dubey, R., Wamba, S.F., Childe, S.J., Hazen, B., Akter, S.: Big data and predictive analytics for supply chain and organizational performance. J. Bus. Res. **70**, 308–317 (2017)
47. Hazen, B.T., Skipper, J.B., Ezell, J.D., Boone, C.A.: Big data and predictive analytics for supply chain sustainability: a theory-driven research agenda. Comput. Ind. Eng. **101**, 592–598 (2016)
48. Fang, K., Jiang, Y., Song, M.: Customer profitability forecasting using big data analytics: a case study of the insurance industry. Comput. Ind. Eng. **101**, 554–564 (2016)
49. Psaraftis, H.N., Wen, M., Kontovas, C.A.: Dynamic vehicle routing problems: three decades and counting. Networks **67**, 3–31 (2016)
50. Kalyvas, C., Kokkos, A., Tzouramanis, T.: A survey of official online sources of high-quality free-of-charge geospatial data for maritime geographic information systems applications, http://www.sciencedirect.com/science/article/pii/S0306437916304185 (2017)
51. Giannopoulos, G.A.: The application of information and communication technologies in transport. http://www.sciencedirect.com/science/article/pii/S0377221703000262 (2004)
52. Musa, A., Gunasekaran, A., Yusuf, Y., Abdelazim, A.: Embedded devices for supply chain applications: towards hardware integration of disparate technologies. Expert Syst. Appl. **41**, 137–155 (2014)
53. Lei, L., Fan, C., Boile, M., Theofanis, S.: Collaborative vs. non-collaborative container-vessel scheduling. Transp. Res. Part E Logist. Transp. Rev. **44**, 504–520 (2008)
54. Ascencio, L.M., González-Ramírez, R.G., Bearzotti, L.A., Smith, N.R., Camacho-Vallejo, J. F.: A collaborative supply chain management system for a maritime port logistics chain. J. Appl. Res. Technol. **12**, 444–458 (2014)
55. Feng, F., Pang, Y., Lodewijks, G.: An intelligent agent-based information integrated platform for hinterland container transport. In Proceedings of 2014 IEEE International Conference on Service Operations and Logistics, and Informatics, pp. 84–89. IEEE (2014)
56. Notteboom, T.E., Rodrigue, J.-P.: Port regionalization: towards a new phase in port development. Marit. Policy Manag. **32**, 297–313 (2005)

Analyzing the Impact of a Decision Support System on a Medium Sized Apparel Company

Ebru Gökalp, Mert Onuralp Gökalp and P. Erhan Eren

Abstract Decision Support Systems (DSS) have been considered as a crucial tool providing competitive advantages, productivity as well as flexibility. The system analysis of a representative medium sized apparel company is performed to derive the requirements of the specific DSS developed to analyze the effects of using it in the apparel company. The observed benefits of the DSS after one year of use are providing efficient use of capacity, accordingly decreasing production costs, effective and accurate calculation of pre-cost estimation, improved user performance and response time to customers, as well as increased financial turnover rate, customer service and decision quality.

Keywords Decision support system · Specific decision support system
Production planning · Pre-cost estimation · Apparel · SME

1 Introduction

Over the past several decades, the thriving textile and apparel industry has been an essential part of Turkey's success story. It has been the engine of employment since the 1950s in Turkey. According to sector shares in terms of contribution to exports, the textile and apparel sector has a share of 11.8% of total exports of Turkey in 2015, with an amount of 25 billion dollars, which is the highest. Automotive sector follows it by ranking the second by 21.2 billion dollar exportations [1]. Overall, Turkey ranks fifth in the world for apparel exports.

Despite the size of the industry, most firms are Small and Medium Sized Enterprises (SMEs), 90% of the export oriented firms in the textile and apparel industry are SMEs in Turkey. The total number of firms in this industry in Turkey is 18,500, 60% of them are in the apparel domain, 40% of them are textile firms. The

E. Gökalp (✉) · M. O. Gökalp · P. E. Eren
Information Systems, Informatics Institute, Middle East Technical University,
Ankara, Turkey
e-mail: egokalp@metu.edu.tr

© Springer International Publishing AG 2018
R. Valencia-García et al. (eds.), *Exploring Intelligent Decision Support Systems*,
Studies in Computational Intelligence 764,
https://doi.org/10.1007/978-3-319-74002-7_4

SMEs have an important role to play in economic development, poverty reduction and employment creation in Turkey. Moreover, apparel production is a high value-added industry in the global textile manufacturing chain. Thus, the scope of our study is determined as the SMEs in the apparel sector.

The apparel industry is characterized by unpredictable demand, quick response times, short product life cycles, and high product variety. Besides, the sector has been under pressure due to global competition and the removal of quotas on China in 2005. Competition in the global world market, especially the competition with firms in China is getting tougher. Under these competitive environments, apparel producers pay more attention to use all resources efficiently, to have a more flexible production system to meet changes in the market and send goods to customers as early as possible to satisfy the increased customer expectation, to utilize workforce effectively, and to access higher productivity levels.

To stay competitive in today's business world with strong globalization and rapid technological changes, apparel companies need to engage in the global market, apply the latest technology and actively structure them to be ahead of their competitors. Investing in information systems (IS) becomes a crucial topic in the industry [2, 3]. It has been thought as a good strategy to improve organizational efficiency and effectiveness and to reduce the costs [4]. IS has been widely considered as an essential tool to provide competitive advantage, productivity as well as flexibility for SMEs [5, 6]. Studies have shown that the ability to apply an appropriate portfolio of IS capabilities is critical to achieve sustainable improvement in competitiveness [7, 8].

SMEs and large firms do have major differences in term of level of resources, structure, internal power conflict, and bargaining power with suppliers [9]. SMEs often have a low level of internal IS competence [10]. In addition, organizational processes are needed to exploit such skills to help them identify and realize IS opportunities in the organizations.

Decision support system (DSS) is described as "*computer based information systems that provide interactive information support to managers during the decision-making process.*" [11]. In general, it supplies many benefits for production planning and marketing department of a manufacturing company, such as decision quality, improved communication, cost reduction, increased productivity, time savings, improved customer and employee satisfaction [12].

The purpose of this study is to analyze the benefits and challenges of an IS, which is classified as specific DSS, for middle management of the production planning and marketing departments of a medium sized representative apparel company. Towards this purpose, the system analysis of the company is performed, the requirements of the DSS to be implemented are derived based on this system analysis, then, the DSS is developed, and finally the benefits and challenges of the DSS are analyzed, where a one-year analysis was carried out in order to identify and compare the improvements achieved by the use of this system in a real life setting. Real data, collected as part of this deployment, is utilized to illustrate the effects of such a DSS.

Since export oriented apparel factories' production is on the make-to-order basis instead of make-to-stock, there is no option like an inventory of goods that can be sold at another time. Consequently, the production planning is complicated and a vital process in the apparel sector. One of the components of the developed DSS is the production planning module which focuses on determining the delivery date of orders. It is designed to use the production capacity efficiently, prevent problems which arise due to idle capacity and capacity shortages. The decision of determining the delivery date of orders is made with the help of the DSS instead of the traditional manual way. This provides a means to decrease mistakes made by the users, to respond to customers in a short time, to use resources effectively and to decrease loss.

Another component of the developed DSS is the pre-cost estimation module which aims to calculate the total cost, which consists of fabric, materials, labor and overhead costs, and the price for customers under different profit margin options in a short amount of time, by using the DSS instead of calculating manually and responding to customers to give information about the price in a short time to be one step ahead of other competitors.

Another purpose of the developed DSS is to integrate information from different departments of the organization and use this information for production planning and pre-cost estimation processes in order to eliminate inconsistent data and workforce. Gathering, storing and processing data help the company to increase its efficiency and effectiveness by improving decision-making process.

The paper is organized as follows: background of the study is given in the next section, followed by the design and development of the proposed DSS. After that, the results are discussed. Finally, remarks about the developed DSS conclude the paper.

2 Background

2.1 Related Works

There is a number of DSS developed for the apparel industry as summarized below:

- Wong and Leung [13] propose a genetic optimized decision-making model for improving the utilization of fabric consumption by developing feasible cutting order plans. However, their solution is restricted to a single production step, fabric cutting.
- Aksoy and Öztürk [14] develop a fuzzy logic-based global outsourcing DSS to evaluate the "make or buy" decision in the apparel industry. Their main focus is the outsourcing aspect in the apparel industry.
- Guo et al. [15, 16] present an intelligent-DSS utilizing radio frequency identification (RFID) technology to handle production tracking and scheduling in a distributed garment manufacturing environment.
- Chen et al. [17] propose an order allocation DSS to address the problem of capacity planning as part of global logistics operations improvement in the case

of order allocation among multiple factories by applying the most suitable order allocation model.

- Jongmuanwai et al. [18] provide profit prediction within the context of a DSS implementation for the garment industry. An approach incorporating goal seek and scenario ("what-if") analysis is utilized to compare the accuracy of their prediction with respect to measured values.

Although there are various DSS developed in the literature for the apparel domain, there are not any studies analyzing the impact of the DSS use on textile and apparel companies for pre-cost estimation and production planning purposes, especially based on real data highlighting the difference such a system makes in the case of a medium sized company utilizing such a sample DSS. Hence, this is the main focus of the study. A specific DSS is developed for the Representative Apparel Company to implement all of the derived requirements including specifics of the company.

2.2 The Representative Apparel Company

For this research, "The Company" was chosen as a representative apparel company in Turkey. The Company, family owned, was founded in 1995. After 3 years from its foundation, it decided to establish other factories located in Mersin Free Zone to increase production capacity. The group is made up of three companies now, additionally other factories are sometimes used as a subcontractor by the company when the capacity is not sufficient for orders. It carries out its activities and operations with more than 800 well-trained, experienced employees working in a totally covered area of 8660 m^2, and specializes in women clothing. In house capacity is 250,000 pieces of shirt a month. Goods are sent mainly to customers living in Germany, Holland, Belgium, and France.

The Company has a functional organizational structure. The six functions which consist of production, marketing, production planning and purchasing, quality management, human resources and finance, and quality assurance departments are directly reporting to the General Manager who is the owner of the company.

Data are collected and stored as both hardcopy and softcopy as MS excel or MS word files in the original setup. Since they do not have any database system, they have to enter and store the same data in different files. Different documents are prepared by using the same data. Inconsistencies are seen among data in different documents. The purpose of the developed DSS is to integrate information from different divisions of the organization in a centralized database.

In the original setup, some problems are faced by The Company such as paying compensation for late delivery. For instance, they had to pay 10,500 € to their customers in 2014 because of late delivery and The Company received many customer complaints about not sending goods on time, the goods had to be sent by plane instead of by truck, in order to compensate for the delay in the production.

This causes customer dissatisfaction and extra delivery cost. Complaints are also about not responding to customer mails on time, when they ask about the product price.

3 The Proposed DSS

3.1 The Design of the DSS

The proposed system suggests a specific DSS to assist the production planning head and the general manager, who is also the marketing department head, in their semi-structured decisions, which refer to the decisions requiring human judgment and at the same time some agreement on the solution method. The DSS is designed to provide support in the following fields of decision making: the pre cost estimation and the production planning.

A detailed analysis of the existing system in the Representative Apparel Company is conducted. As a result of analyzing the workflow and the information flow in detail, the data flow diagram (DFD) at the highest level, given in Fig. 1 below, is generated.

The requirements of the proposed system are as follows:

1. **Record Data**: Customer sends potential order information, and the information flow and the workflow start accordingly. The order information is recorded as order data in the developed DSS. The user can start to use the system and record the data after user authentication is successful, as seen in the main screen given in Fig. 3.
2. **Define Resources**: The purchasing manager is informed about the order as soon as order information is recorded in the system. She finds appropriate fabric and accessories for the order based on the order information, and then records their receiving dates and costs. The CAD-CAM department determines the unit consumption of the fabric based on the model description. These are recorded as fabric and material data in the developed DSS.
3. **Analyze Operations**: The work-study department calculates the standard allowed minutes (SAM) of the order, which refers to the standard time (minutes) allocated to produce one garment. SAM of a product varies according to the work content or simply according to the number of operations, length of seams, fabric types, stitching accuracy needed, and sewing technology to be used. It is calculated based on the sketches of the garment. Then, how many pieces can be produced in a day with the total number of workers in a sewing line, and how many days the production can be completed are calculated. This is recorded as labor data in the DSS.
4. **Determine Optimal Due Date**: Considering the operation analysis information in addition to in-house and subcontractors' capacities, and also fabric and materials delivery date information, the production planning department

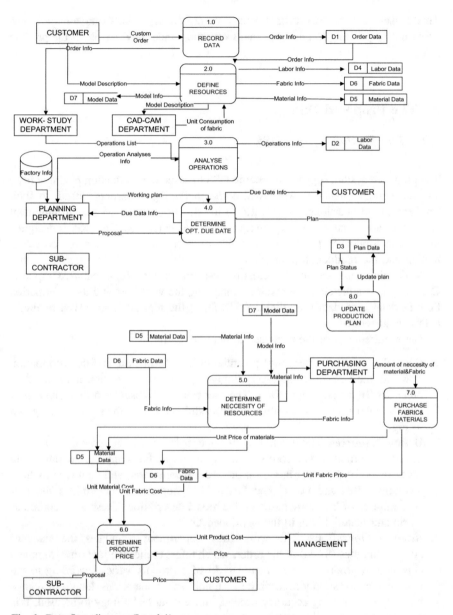

Fig. 1 Data flow diagram (Level 1)

determines the delivery date of the order. Thus, the master production scheduling is done. This is recorded as plan data in the DSS. Because of tables, forms and queries in the system, the production planning department will decide the most appropriate delivery date of an order easily.

5. **Determine Necessity of Resources**: The production planning department starts to control fabric and accessories of the order. The material resource is planned by the production planning department, and the purchasing department supplies necessary items according to this information. As a result of this step, the fabric and material data in the DSS is updated with their unit prices.

6. **Determine Product Price**: As a result of updating the fabric and material data with their unit prices by the purchasing department, the product price is calculated based on the unit cost of material, fabric, and labor costs. If the in-house capacity is not sufficient, proposals are sent to sub-contractors, then they send their unit production price for just CMT (cutting-making-trimming) operations back. The screenshot of the pre-cost estimation module is given in the Fig. 5 below. This screen supports the general manager semi-structured decision of the price to be given to the customer. The general manager decides on the product price based on this information and the customer is informed about the proposed product price and the delivery date of the order. After negotiation, if it is agreed on, the order is added to the plan. Otherwise, it is discarded from the plan.

7. **Purchase Fabric & Materials**: The material resource is planned by the production planning department, and the purchasing department supplies necessary items according to this information after the customer approves the price and the delivery date of the order.

8. **Update Production Plan**: The production planning department decides where to produce and which product line(s) will produce the order. After that, the production planning department starts to follow the order, controls if it can be started when it is planned, if there is a shortage of any accessories, and if they can be delivered on time. The shop floor management is carried out by the managers of the production lines. The order plan is given in Fig. 4 below. After fabric and accessories have arrived, and the planned production time has come, cutting, sewing, washing (optional), final process, packaging, shipping will be done in sequence.

3.2   The Development of the DSS

The system architecture of the proposed DSS comprises modules of User Interface, Database Management, Order Management, Operation Analysis, Reporting, Production Planning, and Pre-Cost Estimation, as shown in Fig. 2.

User Interface (UI) Module
This module manages the interaction between users and the system. It provides input data relating to orders and output information regarding pre-cost estimation and production planning. The main screen of the DSS is given in the Fig. 3.

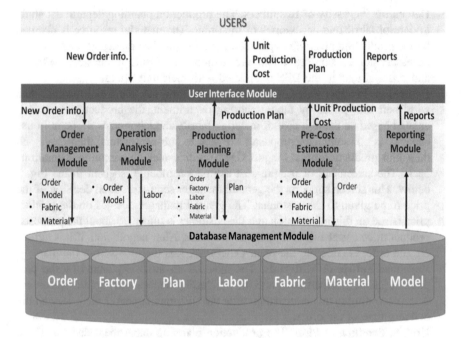

Fig. 2 General architecture of the proposed DSS

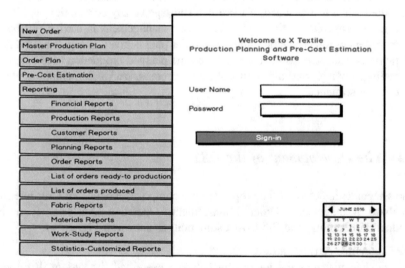

Fig. 3 The main screen

Database Management Module

The DSS inputs, stores, and uses data (order, factory, plan, labor, fabric, material, model) in the centralized database and calculates material requirements, unit production cost, production planning. The Database Management Module provides management of this data for the DSS. It serves as a centralized data store and helps with the data consistency as well as the integration of the modules.

Order Management Module
This module provides the management of the order information as order (quantity, customer, etc.), model (sketch of garment), fabric, material, etc. The functions of "Record Data" and "Define Resources" shown in the DFD in Fig. 1 are performed via this module.

Operation Analysis Module
It aims to provide the unit labor cost calculation of the production based on the sketch of the garment. The function "Analyze Operations" shown in the DFD is performed via this module.

Reporting Module
This module provides reports of financial, production, customer, planning, order, list of orders ready-to production, list of orders produced, fabric, material, work-study as well as customizable reports. Reports which can be generated to support the decisions to be given are shown in the main screen of the DSS given in Fig. 3.

Production Planning Module
The production planning module focuses on determining the delivery date of orders. It is designed to use the production capacity efficiently, prevent problems which arise from idle capacity and capacity shortage.

In the module, the first level of the production planning addresses the long-term aspect of production planning. It determines weekly production quantities as well as overtime and subcontracting levels to minimize the total production cost and determine delivery dates of orders.

The production planning department develops the master production scheduling and determines the delivery date of the order based on the information of the operation analysis information, factories capacities and also fabric and materials delivery dates. The optimal delivery date is determined by using the DSS. An optimization algorithm is used for the production planning module as part of a model-driven approach. Moreover, actual produced quantity and actual production efficiency, which are also stored in the DSS, are used for future master production scheduling. The order plan is shown in Fig. 4. The master production planning includes the capacity fulfillment ratio of the factories for the future weeks.

Schedule Variance (SV %) demonstrates the percentage of how much you're under, over, or exactly on target with your schedule for the current level of completion. A positive Schedule Variance tells you that the order is ahead of the schedule, while a negative Schedule Variance tells you that the order is behind the schedule.

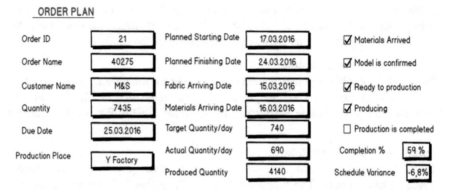

Fig. 4 The order plan

Pre-Cost Estimation Module

The pre-cost estimation module calculates the total cost which consists of fabric, materials, labor and overhead costs and prices for the customers under different profit margin conditions as shown in Fig. 5. The labor cost is calculated and shown to the decision maker for supporting his decision if it is produced in the factory. So, after receiving the order price from the sub-contractor, the decision-maker will decide where to produce.

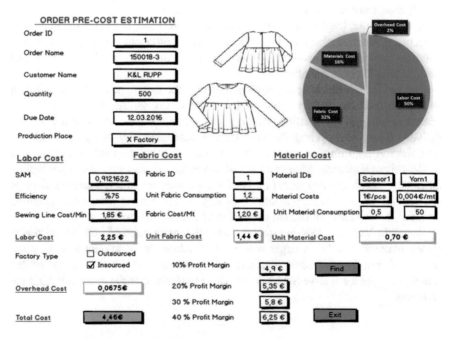

Fig. 5 Pre-cost estimation of the order

Cost estimation is forecasting the cost performed before the order is manufactured. The first step in the costing process can also be called as pre-cost estimation. Pre-cost is a preliminary estimate or "best guess" of what the costs will be to produce the order, based on judgment and past experience. This pre-cost estimation is usually accurate within 10–15% of the actual cost and gives the marketing department some idea for whether to accept the order or not [19]. Summarized cost data, previous garments cost sheets and fabric requirement approximations are used for pre-cost estimation for a proposed new order [20]. The costs in apparel companies are classified based on the components as material costs, labor costs, fabric cost and overhead cost [21].

Overhead cost for each order is calculated as the formula given below. The monthly overhead cost is the summation of the costs of the items including rent, electricity, maintenance, indirect salaries, communication, transport, security, and office equipment.

Some formulas used in the DSS are as follows:

Sewing Line Cost/min

$$= \frac{\text{Total Salary of a sewing line for one day}}{\text{Total no. of Man-power Daily working hours} * 60 * \text{Line efficiency}\,(\%)}$$

$$\text{Labor Cost}\,(\text{€}) = \frac{\text{Standard Allowed Minutes}\,(\text{minute}) * \text{Sewing Line Cost}\,(\text{€/minute})}{\text{Efficiency}\,(\%)}$$

$$\text{Overhead Cost}\,(\text{€}) = \frac{\text{The monthly overhead cost}\,(\text{€/month}) * \text{Labor Cost}\,(\text{€})}{\text{Direct labor cost}(\text{€/month})}$$

$$\text{Fabric Cost}\,(\text{€}) = \text{Fabric Consumption for one piece(meter)} * \text{Unit Fabric Cost}\,(\text{€/meter})$$

$$\text{Materials Cost}(\text{€}) = \sum_{m=1...n} (\text{unit material consumption} * \text{material cost})$$

$$\text{Total Cost}(\text{€}) = \text{Labor Cost} + \text{Fabric Cost} + \text{Material Cost} + \text{Overhead Cost}$$

$$\text{Price Alternative for}\%5\text{profit}(\text{€}) = \text{Total Cost}(\text{€}) * (1 + 0.05)$$

Production Efficiency

$$= \frac{\text{Production Output from the Line} * \text{Standard Allowed Minutes}\,(\text{minute})}{\text{Total Number of Operators} * \text{Total Working Hours} * 60}$$

$$\text{Schedule Variance}\,(\text{SV}\%) = \frac{\text{Actual Produced Quantity} - \text{Target Quantity}}{\text{Budgeted Quantity of Work Scheduled}}$$

4   Results and Discussion

The DSS is designed to overcome the drawbacks of the original setup. The Company has started to employ the developed DSS since the beginning of 2015. The observed benefits of the DSS after one year of use in The Company compared to the previous year are summarized in Table 1.

- **Decreased Penalty Cost**: After 2015, The Company has not received any penalty for sending any orders late. The penalty cost for late delivery received in 2014 is 10,500 €.
- **Increased Financial Turnover Rate**: The financial turnover rate has increased 18% with the same workforce capacity compared to 2014. There are many parameters that affect this value. The capacity is used more efficiently, unnecessary setup costs are removed, idle capacity and capacity shortage problems are solved. Accordingly, the cost of production is reduced.
- **Decreased Overtime Rate**: The overtime rate is reduced in the company by 75% after 2015. Overtime can be used to adjust the capacity to handle the jobs which otherwise become tardy and may lead to poor reputation or in the worst case, losing valuable customers.
- **Improved Production Quality**: Since production manager is not dealing with sewing line scheduling anymore, he just concentrates on balancing the load on the line and training the operators. That improves the produced garment quantity and quality. There is no recorded customer complaint about the quality of the products in 2015.
- **Improved Decision Making Process**: Prior to employing the DSS, the production department made decisions, such as the confirmed due date by using multiple excel files and computations. This manual approach required too much effort. However, the time required to find out the optimum solution in the DSS

Table 1 The observed impacts of the developed DSS

Parameter	2014	2015	Change
Amount of penalty for sending the orders late	€10,500	€0	100% decrease
Financial turnover rate	€34,300,000	€40,474,000	18% increase
Amount of overtime	3560 man-hour/ month	890 man-hour/ month	75% decrease
# of customer complaint about the quality of the products	4	0	100% decrease
Average productivity rate of sewing lines	75%	82%	9.3% increase
Required time for calculating of the optimum due date of a new order (This affects average response time to customer)	38.5 min	2.25 min	94% decrease

after entering order information is very short. Therefore, the production department team has immediate access to the results (~ 2.25 min) that are stored on the DSS. The DSS simplifies the work of the production department, and allows it to make decisions in less time.

- **Improved Decision Quality**: The impact of the DSS to the business has been significant in several ways. Primarily, the DSS is an important tool for the production planning team. It substantially decreases the team's decision-making time and provides it with better and more detailed information than it could obtain with a manual solution.

- **Improved Customer Satisfaction**: The DSS also allows the team to use the capacity more efficiently and be more responsive in informing the customer about the inappropriateness of a proposed due date, which in turn increases customer satisfaction.

- **Improved Productivity**: Production scheduling is performed by using the DSS rather than traditional determination of delivery date based on experience. This prevents problems of idle capacity and shortage of capacity, which are the major drawbacks of the original setup. The decision maker can forecast if there would be a capacity shortage or idle capacity of the factories based on the capacity fulfilment ratio given in the master production planning, and he has time to prepare an action plan to prevent these kinds of problems.

- **Improved Accuracy and Consistency of Data**: The centralized database, designed for effective integration of information, provides improved communication, consistent data, correct calculations for generating reliable reports, and accurate system for calculating the pre-cost estimation.

- **Improved User Performance**: Since data in different files are gathered together and integrated in a well-organized way, the proposed system reduces the efforts of organized documentation. Besides, data are entered in the system easily; the required production performances are obtained from the system in a fast and reliable manner.

5 Conclusion

ISs can help apparel SMEs create business opportunities as well as combat pressure from competition. Appropriate IS provides benefits for such SMEs in cutting costs by improving their internal processes and improving their business through faster communication with their customers. In fact, IS has the potential to improve the core business of apparel SMEs in every step of the business processes.

Throughout the study, it is intended to analyze the benefits and challenges of an IS, which is classified as specific DSS, for a representative apparel company. Towards this purpose, the DSS is designed and developed to support decision making of the pre cost estimation and production planning. Then, the DSS is implemented and started to be used, and the benefits and challenges of the DSS are analyzed after its use for one year in the representative company. It is observed that the developed DSS provides improvements in decision making, financial turnover rate, production quality, decision quality, customer satisfaction, productivity, user performance, data accuracy and data consistency, as well as decreases in penalty cost and overtime rate.

The contribution areas of the study include:

- The requirements associated with such a generic DSS for the textile and apparel industry have been analyzed and presented in conjunction with a constructed Data Flow Diagram.
- The DSS tailored specifically for a medium sized company has been developed according to these identified characteristics.
- By deploying this DSS in a medium sized apparel company, the benefits have been analyzed and highlighted, where a one-year analysis was carried out in order to identify and compare the improvements achieved by the use of this system in a real life setting. Real data collected as part of this deployment is utilized to illustrate the effects of such a DSS.

On the other hand, there are some limitations of the developed DSS. The future work to address these limitations is summarized as follows:

- For the sewing process, an optimization is planned to be designed in order to solve the line balancing problem which is done by the manager of the line with his own experience. Thus, a scientific method is planned to be used instead of the traditional one.
- The boundaries of the system are planned to be extended. The system is used only by the planning, purchasing and marketing departments. The reason is the use of an evolutionary approach, which is an iterative and incremental way of developing the DSS. Since production planning and pre-cost estimation are the most critical activities for the company, as a starting point, these functions are selected as core of the DSS. For the second incremental iteration, the DSS is planned to be evolved by adding support for decision making in the field of customer relationship management.
- The planning, purchasing and marketing departments were willing to use such a specific DSS, the end-user training took a short of time and they did not resist to use the DSS. However, the user interfaces (UI) of the DSS are planned to be improved based the usability standards in order to make it easier-to use before extending the boundaries of the system.

Acknowledgements Our profound thanks go to "The Company" that participated in this study. We would like to thank Prof. Dr. D. Tayyar Şen for his suggestions and inexhaustible support.

References

1. TEA: http://www.tim.org.tr/ (2017)
2. Lai, F., Zhao, X., Wang, Q.: The impact of information technology on the competitive advantage of logistics firms in China. Ind. Manag. Data Syst. **106**, 1249–1271 (2006)
3. Nie, J.: A study of information technology adoption for small and medium sized enterprises strategic competitiveness. In: Wireless Communications, Networking and Mobile Computing, 2007. WiCom 2007. International Conference on. pp. 4342–4346. IEEE (2007)
4. Al-Mashari, M., Zairi, M.: Supply-chain re-engineering using enterprise resource planning (ERP) systems: an analysis of a SAP R/3 implementation case. Int. J. Phys. Distrib. Logist. Manag. **30**, 296–313 (2000)
5. Bertolini, M., Bevilacqua, M., Bottani, E., Rizzi, A.: Requirements of an ERP enterprise modeller for optimally managing the fashion industry supply chain. J. Enterp. Inf. Manag. **17**, 180–190 (2004)
6. Moghavvemi, S., Hakimian, F., Feissal, T., Faziharudean, T.M.: Competitive advantages through IT innovation adoption by SMEs (2012)
7. Jung, J.H., Schneider, C., Valacich, J.: Enhancing the motivational affordance of information systems: the effects of real-time performance feedback and goal setting in group collaboration environments. Manag. Sci. **56**, 724–742 (2010)
8. Doherty, N.F., Terry, M.: The role of IS capabilities in delivering sustainable improvements to competitive positioning. J. Strateg. Inf. Syst. **18**, 100–116 (2009)
9. Cragg, P., Caldeira, M., Ward, J.: Organizational information systems competences in small and medium-sized enterprises. Inf. Manage. **48**, 353–363 (2011)
10. Hashim, J.: Information communication technology (ICT) adoption among SME owners in Malaysia. Int. J. Bus. Inf. **2**, (2015)
11. O'Brien, J.A., Marakas, G.: Introduction to information systems. McGraw-Hill, Inc. (2005)
12. Aronson, J.E., Liang, T.-P., Turban, E.: Decision support systems and intelligent systems. Pearson Prentice-Hall (2005)
13. Wong, W.K., Leung, S.Y.S.: Genetic optimization of fabric utilization in apparel manufacturing. Int. J. Prod. Econ. **114**, 376–387 (2008)
14. Aksoy, A., Öztürk, N.: Design of an intelligent decision support system for global outsourcing decisions in the apparel industry. J. Text. Inst. **107**, 1322–1335 (2016)
15. Guo, Z.X., Ngai, E.W.T., Yang, C., Liang, X.: An RFID-based intelligent decision support system architecture for production monitoring and scheduling in a distributed manufacturing environment. Int. J. Prod. Econ. **159**, 16–28 (2015)
16. Guo, Z.X., Wong, W.K., Guo, C.: A cloud-based intelligent decision-making system for order tracking and allocation in apparel manufacturing. Int. J. Prod. Res. **52**, 1100–1115 (2014)
17. Chen, M.-K., Wang, Y.-H., Hung, T.-Y.: Establishing an order allocation decision support system via learning curve model for apparel logistics. J. Ind. Prod. Eng. **31**, 274–285 (2014)
18. Jongmuanwai, B., Angskun, J., Angskun, T.: DSG: A Decision Support System for Garment Industry (2010)
19. Brown, P., Rice, J.: Ready-to-wear apparel analysis. Pearson Higher Ed (2013)
20. Rosenau, J.A., Wilson, D.L.: Apparel merchandising: the line starts here. A&C Black (2014)
21. Horngren, C.T., Foster, G., Datar, S.M., Rajan, M., Ittner, C., Baldwin, A.A.: Cost accounting: a managerial emphasis. Issues Account. Educ. **25**, 789–790 (2010)

A Multicriteria Decision Support System Framework for Computer Selection

Jorge Luis García-Alcaraz, Valeria Martínez-Loya,
Roberto Díaz-Reza, Liliana Avelar Sosa
and Ismael Canales Valdiviezo

Abstract Nowadays, buying a computer for family use is a frequent practice, yet the wide variety of equipment that markets offer can be overwhelming. Each computer has its own characteristics and attributes, and some of such attributes—especially qualitative features—may be difficult to assess. This chapter presents a theoretical framework that allows families to evaluate computers from a multi-attribute perspective by using two techniques: the Analytic Hierarchy Process (AHP) and the Technique for Order of Preference by Similarity to Ideal Solution (TOPSIS). The former is used to weight the attributes, whereas the latter is used to propose a solution. A case study is presented to illustrate the computer selection process performed by a four-member family on four alternatives by taking into account four quantitative attributes—cost, processor speed, RAM, and hard drive capacity—and two qualitative attributes—brand prestige and after-sales service. Our findings demonstrate that our AHP-TOPSIS approach is friendly to users, especially to non-expert users, since they can perform the evaluation process on their own.

Keywords Computer selection · AHP · TOPSIS

1 Introduction

Modern technologies have gained importance because they provide a lot of benefits. Their applications range from complex computer systems used in scientific research [1] to much simpler systems for leisure activities, such as video games [2, 3]. In this sense, education is one of the fields most benefitted by the adoption of technology,

J. L. García-Alcaraz (✉) · V. Martínez-Loya · L. A. Sosa
Department of Industrial Engineering and Manufacturing, Universidad Autónoma de Ciudad Juárez, Av. del Charro 450 Norte, Ciudad Juárez 32310, Chihuahua, Mexico
e-mail: jorge.garcia@uacj.mx

R. Díaz-Reza · I. C. Valdiviezo
Department of Electric and Computational Sciences, Universidad Autónoma de Ciudad Juárez, Av. del Charro 450, Ciudad Juárez 32310, Chihuahua, Mexico

© Springer International Publishing AG 2018
R. Valencia-García et al. (eds.), *Exploring Intelligent Decision Support Systems*,
Studies in Computational Intelligence 764,
https://doi.org/10.1007/978-3-319-74002-7_5

since such technological advancements have defied traditional teaching and learning methods. For instance, laptops have enhanced learning by encouraging different learning and work strategies, such as teamwork and real-time communication. Likewise, learners have the opportunity to instantly explore the wide range of resources available for them on the Web [4].

1.1 Computer Adoption

Many aspects contribute to the adoption of computers. In the past, learners had to take many books to classes; nowadays, many school activities can be performed using only a laptop, since such computers can store not only educational content (e.g. books, tasks, notes.), but also any other type of information (e.g. photos, videos, music). Moreover, laptops are light and easy to carry and manipulate. Also, current financial trends provide almost all the sectors with the opportunity to purchase a computer.

In 2012, Mexico reported 42.3 million computers with access to educational information on the Internet, as 36.7 million people had access to this service. However, such numbers increased in 2015, when 55.7 million people had a computer and 62.4 million users had internet access via computers or mobile devices (INEGI, 2015). Such data may seem high, yet when comparing Mexico with other regions, we can observe a significant difference. For instance, in 2014, over 91% of households in the United Kingdom and Germany had at least one computer [5], and in 2016, the worldwide personal computer sales report reached a total of 269.16 million units. Also, for the first semester of 2017, around 123.29 million units have been sold [6].

1.2 Computer Attributes

Nowadays, each computer brand offers equipment for distinct uses and a set of characteristics or attributes that differentiate it from other brands (e.g. cost, versions, and designs). Similarly, customers usually have an overall idea of what they want to purchase, yet the wide range of alternatives can make it difficult to choose the correct computer, no matter how clear consumer expectations are. In general, purchasing computer equipment cannot be taken lightly, since the final decision can affect a family's budget and investment capabilities. Therefore, before purchasing a computer, it is important to take into account the following aspects:

- Choosing between desktop or laptop computer.
- Identifying the computer usage.
- Reviewing customer opinions about popular brands.
- Analyzing the hard disk type and capacity.
- Identifying RAM options as well as the capacity for future expansion.
- Defining processor speed and motherboard type.

- Selecting the ideal operating system according to its purpose.
- Considering battery life.
- Considering screen size.
- Considering the characteristics of graphics and sound cards.
- Considering the number of ports, expansion slots, and peripherals that can be connected.
- Considering the warranty period.

1.3   The Multi-attribute Approach for Alternative Selection

As mentioned above, buying a computer implies simultaneously assessing the viability of a set of attributes, which implies that a decision based merely on costs is no longer enough. In the past, techniques such as Internal Rate of Return (IRR), Present Value (PV), Future Value (PV), Equivalent annual cost (EAC), and Return Period were proposed as means to perform evaluations from an economic approach [7]. Nevertheless, such techniques were hardly criticized for integrating only financial or quantifiable aspects. As a result, qualitative attributes of computer equipment, such as processor speed and memory capacity, have been usually neglected because current evaluation processes are unable to successfully integrate them in the assessments. Moreover, economic-based or financial-based decisions on computer selection are easier and cheap to make, since they can be delegated to a single person [8].

To address the limitation of quantitative evaluations, multi-attribute techniques have been proposed to evaluate several attributes simultaneously. In this sense, Fig. 1 below depicts a brief classification of these multi-attribute techniques. Also, the figure shows the attributes considered by each technique when proposing a solution. As can be observed, the multi-attribute techniques of the first category analyze a simple objective to make the decision and require deterministic or probabilistic information [9]. On the other hand, the second category refers to decision support systems. Such systems are considered as an evolved part of all the existing decision-making techniques, since they integrate traditional single and multiple-object decision making techniques and methods but are integrated in computer systems that facilitate analysis and avoid errors. Finally, the third category of decision making techniques includes those tools that integrate several objectives and attributes simultaneously. In this research, we focus on this group of techniques, since the problem of computer selection fits into this category [7].

Even though experts have proposed many classifications for decision making techniques, there is no a generalized agreement on their taxonomy. That said, this chapter relies on the proposal depicted in Fig. 1 to address the problem of computer selection from a multi-attribute perspective. Also, the applications and benefits of multi-attribute decision-making techniques have been widely reported in the literature. According to [10] some of such benefits are the following:

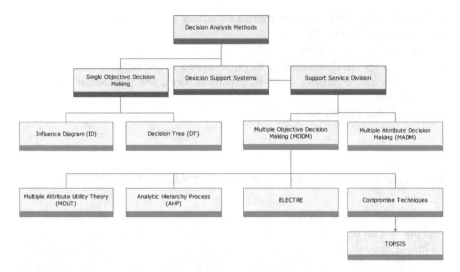

Fig. 1 Multicriteria decision-making techniques

- These techniques allow the simultaneous evaluation of several attributes, which can be quantitative and qualitative, since not only price should be evaluated, but also operating system type and connectivity, among others.
- A decision group can be integrated to conduct the evaluation of the alternatives, thereby increasing the democratic sense of acceptance of the decision, since all the members perform their judgments.
- A different level of importance can be assigned to each attribute, giving greater importance to those attributes that are considered more important for the group.
- Different objectives can be evaluated simultaneously, since some attributes—such as cost—are to be minimized, while others (e.g. hard disk storage capacity or RAM) must be maximized.

1.4 Research Problem and Objective

As mentioned above, each computer has its distinctive attributes. Although these factures have evolved over time, the process of choosing the best computer equipment or alternative remains complicated and complex, as it requires a simultaneous, multi-criteria analysis. Moreover, preferences in attributes such as costs, brand, technical specifications [11], accessories, and warranties, and even the opinions of family and friends can influence or hinder the computer acquisition process. Moreover, it seems that current decision-making frameworks and proposals are unable to simplify the increasingly necessary and common task of computer selection among households. To address such limitations, we propose a multi-attribute and multi-criteria evaluation approach whose goal is to support computer selection among households by considering their needs and financial capabilities. This approach relies on TOPSIS and AHP to integrate the various computer attributes—both quantitative and qualitative. Also, the decision group is integrated by a family who plans to purchase a computer. Finally, our approach seeks to complete three tasks: to make a final decision through family consensus, to acquire normal, not specialized, software, and to evaluate several attributes with different levels of importance.

2 Methodology

Figure 2 illustrates the activities and steps completed to perform the computer evaluation and selection process. As previously mentioned, this selection and evaluation approach relies on multi-attribute techniques, which simultaneously evaluate a set of attributes and include the participation of all the people involved in the decision-making problem.

2.1 Integrating the Decision Group

Multi-criteria and multi-attribute decision making approaches demand the involvement of all the group-decision members. Therefore, this study takes into account the active participation of all the family members involved in the computer selection process, since the level of importance of each qualitative attribute may vary across

Fig. 2 Methodology for computer selection

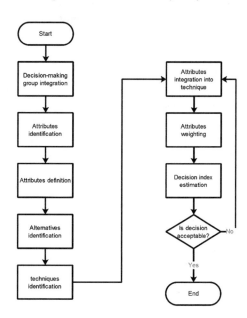

such members. Thus, in this study family members integrate the decision group hereinafter referred to as E_1, E_2, ... E_n, where n refers to the number of people in the decision group. Also, note that this model assumes that the n family members involved in the decision-making process have the same level of importance. That is, the family's hierarchy does not interfere in the decision that the group makes.

2.2 Identifying Attributes

To identify the computer attributes that are key to computer selection and evaluation, we conducted a literature review on databases such as ScienceDirect, EBSCO Host, and Ingenta using keywords such as *computer selection* and *computer evaluation*. Table 1 shows the attributes identified as decisive features when selecting a computer, mainly a desktop computer, although some of them are also applicable to laptops.

As can be observed, some key computer attributes are quantitative, whereas some others are qualitative. Quantitative attributes or features can be represented by a unit of objective measurement, such as the amount of RAM or hard disk capacity, but qualitative attributes are subjective and need to be selected by a group of experts [12, 13].

2.3 Selecting Evaluation Attributes

Once the decision group is integrated and the key attributes are identified, we must determine which of such attributes need to be evaluated. That said, according to the

Table 1 Classification of key computer attributes

Qualitative	Quantitative
Bluetooth enabled	Cache memory
Brand name reputation	Graphics card
Color	Hard drive capacity and type
Enclosure type	Media card reader
Energy star certification	Number of ports (dvi, Ethernet, HDMI, USB)
Speaker(s) included	Number of internal 3.5″ bays
Warranty from manufacturer and seller	Number of pci-e slots
Wireless networking (display, keyboard, mouse)	Number of ps/2 ports
Operating system	Physical dimensions (height, width, depth, weight)
Optical drive type	Processor speed
Processor brand and model	Memory RAM speed and expansion
Screen multi-touch	Screen resolution and size
	Video memory size

multi-attribute theory, attribute-based evaluations should ideally include less than nine attributes, since from a psychological perspective, evaluations with more than nine attributes are unreliable. However, in this research we require a sufficient number of attributes to adequately represent all the alternatives' characteristics.

2.4 Identifying Alternatives

After selecting the attributes to be evaluated, the decision group must identify the alternatives (i.e. computers) available in the market. This is probably the longest and most effort-consuming step, as it involves exploring every potential alternative currently available. To support this task, experts recommend the following:

- Visit stores specialized in computer equipment sales.
- Browse on the website of online stores dedicated to computer sales and identify the alternatives that could be purchased. Also, check shipping fees, taxes, and any additional transportation costs. For comparative purposes, all additional costs are considered for each alternative as part of its final purchase amount.
- Fairly define computer use and necessary software, since the specifications and attributes to be evaluated change radically.
- Analyze aspects associated with future expansion capacity, such as RAM expansion capacity.
- Identify customer opinions/reviews regarding after sales service and technical assistance.

2.5 Selecting a Multi-attribute Technique

The next step is to identify the best decision-making technique. Many multi-criteria evaluation techniques are used in decision-making processes that involve two or more evaluation attributes. Figure 1 illustrates a broad classification of multi-criteria decision-making techniques. That said, due to their usefulness in alternative assessment processes, we relied on both AHP and TOPSIS to evaluate the best family computer.

AHP has been effectively employed to select the best facilities location [14, 12], to measure the levels of utilization of advanced manufacturing technologies [15], to choose advanced manufacturing technologies [16] to measure the efficiency of lean manufacturing tools [17], and even to decide on the proper location of storage warehouses [12]. On the other hand, TOPSIS has been useful in evaluating company performance [18], selecting better suppliers [19], and evaluating the ergonomic compatibility of advanced manufacturing systems [20] and their utilization levels [15].

As regards the TOPSIS-AHP tool as a hybrid methodology, it has been reported in bibliometric evaluation [21] and water treatment system selection [22] works. Also, it has been applied to improve the human resources selection processes [23] and analyze the feasibility of solar energy generating systems in Indian farms [24].

2.6 AHP and TOPSIS Techniques

The following subsections thoroughly address AHP and TOPSIS as they are employed in this work. However, to consult further information on the AHP technique, please refer to [25–28]. Similarly, readers can consult the works of [29–32] for additional information regarding TOPSIS foundations and applications.

2.6.1 AHP: Attribute Weighting

AHP was proposed by Saaty to weight attributes in evaluations. In other words, AHP determines the level of importance of every evaluation attribute; however, some of these features may be in conflict with one another [15], which implies that decision makers must be prepared to sacrifice one or some attributes to gain a much more significant one, based on the decision group's needs. Also, AHP is based on paired comparisons placed in an array as shown in Eq. (1), where attributes are placed in rows and columns in the same order as they appear using the scale shown in Table 2. This scale is used to determine the preference levels of one attribute over another. To this end, a verbal scale is transformed into a numerical scale. Logically, since the same attribute is compared diagonally, the trace of the matrix is a set of ones. If a value holds position "a_{ij}", then position "a_{ij}" presents an inverse value [33].

$$\begin{bmatrix} 1 & \cdots & a_{1j} \\ \vdots & \ddots & \vdots \\ a_{j1} & \cdots & 1 \end{bmatrix} \tag{1}$$

Table 2 Scale proposed by Saaty

Scale	Verbal expression
1	Equal importance of attributes
3	Moderate importance of one attribute over another
5	Strong importance
7	Very strong importance
9	Extreme importance
2, 4, 6, 8	Intermediate values

Table 3 Random indices

n	3	4	5	6	7	8	9	10
RI	0.58	0.9	1.12	1.24	1.32	1.41	1.45	1.49

The pairwise comparison matrix is a positive reciprocal matrix; thus, eigenvalues and eigenvectors are used to estimate the weights. Also, the pairwise comparison matrix shows the levels of consistency of a decision maker [34]. Specifically, the consistency index (CI) of a decision maker is estimated by using Eq. (2), where λ_{max} represents the maximum eigenvalue of the pair-wise comparison matrix, and n stands for the number of attributes evaluated.

$$CI = \frac{\lambda_{max} - n}{n - 1} \tag{2}$$

Based on CI, we can estimate the consistency ratio (CR) to determine the level of consistency of each decision maker by using Eq. (3).

$$CR = \frac{CI}{RI} \tag{3}$$

CR is an average CI generated from positive reciprocal matrices and a random index (RI), illustrated in Table 3. If a decision maker generates an array of paired comparisons with a CR less than 0.10, it means that he has more than 10% of random errors. Therefore, the decision maker must reissue their judgments until they obtain a lower value [35].

Note that in this work, all the family members belonging to the decision group must provide their judgments and generate their own pairwise comparison matrix according to their preference levels. In this sense, Eq. (4) must be used to unify all the judgments in a single matrix [36, 37]. That matrix is used for obtain the eigenvalues and weight for every attribute.

$$a_{ijT} = \left(a_{ij1} * a_{ij2} * \ldots a_{ijn}\right)^{\frac{1}{n}} \tag{4}$$

2.6.2 TOPSIS

Once all the attributes are weighted, the alternatives need to be evaluated with respect to the attributes. To this end, TOPSIS must be employed. TOPSIS determines the Euclidean distance between an ideal alternative and a non-ideal alternative in relation to all the alternatives; thus, attributes and alternatives are considered as vectors in a multidimensional space. Likewise, the goal of TOPSIS is to choose the alternative that shows the shortest distance from the ideal alternative, and simultaneously, the one that has the greatest distance from the non-ideal alternative. However, this is a weighted distance, since the attributes hold different levels of importance [38].

The ideal alternative is represented by $\mathbf{A^+}$ and is integrated by the highest nominal values of the attributes. On the other hand, the non-ideal alternative $\mathbf{A^-}$ is composed of the worst nominal values of the attributes [39]. For instance, the best nominal value for the cost attribute would be the lowest value among of the alternatives, whereas the worst nominal value would be the highest price, since families prefer a low-cost computer. As for the computer's memory (RAM), the best nominal value refers to the higher value of the alternatives, but the worst is the lowest value. In this sense, TOPSIS is often considered as a compromise ranking technique, as it evaluates the two types of attributes: those that need to be maximized and those that must be minimized [40].

In the end, we evaluated a total of k computers, which are represented by $\mathbf{A^k}$ as shown in Eq. (5). Similarly, we assumed that a total of N attributes were evaluated, which are represented by $\mathbf{X_n}$, as Eq. (6) illustrates [41, 30].

$$A^k = (x_1^k \cdots\cdots x_n^k) \quad \text{for } k = 1, 2, \ldots, k \tag{5}$$

$$X_n = (x_n^1 \cdots\cdots x_n^k) \quad \text{for } n = 1, 2, \ldots, n \tag{6}$$

Since TOPSIS relies on the distances of the ideal alternative and the non-ideal alternative, these are symbolically represented according to Eqs. (7) and (8), respectively. At this point, it is important to mention that these two alternatives do not exist in real life; they are rather generated from the data available for each of the evaluated alternatives.

$$A^+ = (x_1^+, x_2^+, \cdots\cdots x_n^+) \tag{7}$$

$$A^- = (x_1^-, x_2^-, \cdots\cdots x_n^-) \tag{8}$$

When estimating the distances between the different alternatives with respect to the ideal solution and the non-ideal solution, some of the evaluated attributes can be expressed in different measurement scales. In such cases, it is necessary nondimensionalize every attribute to perform any operation with vectors. Such nondimensionalization process is performed as illustrated in Eq. (9) [42],

$$TX_n = \frac{X_n}{\|X_n\|} = \left(\frac{x_n^1}{\|X_n\|}, \cdots\cdots \frac{x_n^k}{\|X_n\|} \right) \tag{9}$$

where $\|X_n\|$ represents the Euclidean norm or distance to the origin, which can be obtained according to Eq. (10).

$$|X_n| = \sqrt{\sum_1^x x_i^2} \qquad (10)$$

Note that Eq. (9) is used to normalize each attribute. However, shorter methods are generally used to perform the normalization process directly in the alternatives, as illustrated in Eq. (11), for the ideal alternative, as shown in Eq. (12), and for the non-ideal alternative, as seen in Eq. (13).

$$TA^k = (t_1^k, \cdots \cdots t_n^k) = \left(\frac{x_1^k}{\|X_1\|}, \cdots \cdots \frac{x_n^k}{\|X_n\|} \right) \qquad (11)$$

$$TA^+ = (t_1^+, \cdots \cdots t_n^+) = \left(\frac{x_1^+}{\|X_1\|}, \cdots \cdots \frac{x_n^+}{\|X_n\|} \right) \qquad (12)$$

$$TA^- = (t_1^-, \cdots \cdots t_n^-) = \left(\frac{x_1^-}{\|X_1\|}, \cdots \cdots \frac{x_n^-}{\|X_n\|} \right) \qquad (13)$$

Once all the attributes are assigned to dimensionless values, it is possible to estimate the distance of each alternative under evaluation to both alternatives, the ideal solution and the non-ideal ideal solution. To this end, we employ Eqs. (14) and (15).

$$\rho(A^k, A^+) = \|w * (TA^k - TA^+)\| = \sqrt{\sum_{n=1}^N w_1 * (t_1^k, \cdots \cdots t_n^+)^2} \qquad (14)$$

$$\rho(A^k, A^-) = \|w * (TA^k - TA^-)\| = \sqrt{\sum_{n=1}^N w_1 * (t_1^k, \cdots \cdots t_n^-)^2} \qquad (15)$$

Note that Eqs. (14) and (15) show a w factor to represent the vector of the weights obtained through AHP from the decision group (i.e. family members). Also, because the goal of TOPSIS is to choose the alternative with the shortest distance to the ideal alternative and but the longest distance to the non-ideal alternative, Eq. (16) is used to estimate an index that ranks the preference levels of the alternatives, selecting the one with the lowest index [43].

$$RC(A^+, A^-) = \frac{\rho(A^k, A^+)}{\rho(A^k, A^+) + \rho(A^k, A^-)} \qquad (16)$$

3 Case Study

The objective of this chapter is to propose an AHP- and TOPSIS-based theoretical framework for computer selection. To demonstrate the feasibility of our approach, this section presents a case study that exemplifies the tasks completed to successfully select the computer that best meets the requirements of a family (expressed by the attributes).

3.1 Integrating the Decision Group

A family wants to purchase a computer that would be used at home by all the family members. This family is composed of:

- The father (E_1): An electrical engineer working in a manufacturing industry.
- The mother (E_2): A biomedical engineer who works in a design company.
- Two college students: The son (E_3) is an industrial design engineering student, and the daughter (E_4) is an industrial engineering student.

Also, as mentioned earlier, all the family members are considered to have the same power of decision regarding the purchase of a computer.

3.2 Identifying Alternatives

As a decision group, the family identifies a set of potential computers that can be acquired to meet their needs. After searching in both online and physical specialized stores, the family selects four computers as alternatives. Although each family member uses the computer for a different purpose, the family expects purchased computer to meet the following requirements.

- Not a too high total cost, since the computer is only for domestic use.
- Extensive storage capacity, since the family stores large amounts of data in the form of photos, videos, and documents.
- The selected computer must be able to connect to the Internet and execute different programs used at school, including the Office suites (Microsoft Office, Open Office, etc.), design programs (AutoCAD, SolidWorks, CorelDraw, etc.), and statistical packages (Minitab, SPSS, etc.).
- Enough RAM capacity for multitasking (e.g. checking email while listening to music and running programs).
- Solid processor capable of performing basic tasks, such as running programs and applications continuously, surfing on the Internet, watching videos, and listening to music. In other words, the family looks for a convenient relationship between information-processing speed and the equipment sale's price.

That said, the chosen alternatives refer to computers of distinct brands and models. For respect and privacy purposes, such details are omitted, and the alternatives were named as A_1, A_2, A_3, and A_4.

3.3 Evaluated Attributes

The family as a decision group has decided to consider the following attributes before acquiring the computer. Such attributes may vary across households.

1. Cost (C, $). It refers to the amount payable for the computer and includes a two-year warranty. This is a quantitative attribute that must be minimized. It is expressed in US dollars ($) and represented by C. This attribute is selected because of the family's low income.
2. RAM (RM, GB). It refers to the computer's random-access memory capacity. This attribute is represented by RM and is expressed in gigabytes (GB). It is a quantitative attribute and must be maximized. The family selected this attribute, because the computer must execute different programs simultaneously.
3. Processor speed (PS, GHz). This attribute refers to the number of operations the computer can perform and the speed at which the processor can work. It is a quantitative attribute represented by PS, but it is expressed in Giga Hertz (GHz). The family chose this attribute since they expect the selected computer to work fast.
4. Hard Disk (HD, TB). This attribute refers to the capacity of the hard disk to store information expressed in Terabytes. It is a quantitative attribute whose maximum values are desired, since the family's stored photographs and high-quality designs might occupy a lot space.
5. Brand prestige (P). It refers to the prestige of the brand in the market given the quality of its products. This is a qualitative attribute that requires the opinion of the decision group members. Maximum values in this attribute are desired.
6. After-sales service (ASS). This is a qualitative attribute and indicates the decision makers' judgements regarding the quality of service received after purchasing the computer. After sale services include warranties and technical support service, among others. Maximum values are desired. Also, this attribute was assessed by considering customer reviews.

3.3.1 Quantitative Attributes

As seen in the previous section, four of the selected attributes are quantitative and two are qualitative. Also, one attribute must be minimized, but five need to be maximized. Once the attributes have been identified, we determine their magnitude in each of the alternatives. In this case, the estimates are calculated separately for

Table 4 Quantitative attribute values

A^k	C ($)	RM (GB)	PS (GHZ)	HD (TB)
A^1	289.99	4	2.3	1
A^2	379.99	4	3	1
A^3	349.99	8	2.4	1
A^4	279.99	4	1.6	0.5

the two types of attributes. Table 4 illustrates the characteristics of the alternatives with respect to their quantitative attributes.

3.3.2 Qualitative Attributes

Table 5 illustrates the judgements from the four family members on the two qualitative attributes. The judgements are expressed in values ranging from one to nine, where one indicates the absence of an attribute in a given alternative, and nine indicates a definite presence. As previously mentioned, each family member or decision maker issues their judgment for each alternative. The average judgement for each alternative is shown in the last column, which is used to generate the final decision matrix.

3.3.3 Final Decision Matrix

Table 6 shows the final decision matrix used to analyze the alternatives by using TOPSIS. In this case, a row called optimization is added immediately after the four alternatives, which indicates the ideal meaning of the attributes. Also, as previously mentioned, only one attribute must be minimized and the other five need to be maximized. In addition, considering the attributes optimization (maximize or minimize), two additional rows have been added in Table 6 to represent the ideal alternative and the non-ideal alternative, respectively.

Ideally, the best option is to buy a computer for 279.99 US dollars, with 8 GB of RAM, with a 1 TB hard disk, a processor speed of 3 GHz, a prestige score of 7.75, and an after-sales service score of 8.5 on a scale from 1 to 9. On the other hand, the worst scenario would be buying a computer for 379.99 US dollars that has only

Table 5 Qualitative attribute values

Ak	E1		E2		E3		E4		Average	
	P	ASS	P	ASS	P	ASS	P	ASS	P	ASS
A^1	7	9	7	8	8	9	8	8	7.5	8.5
A^2	8	8	9	9	7	8	5	9	7.25	8.5
A^3	9	6	6	7	7	9	9	8	7.75	7.5
A^4	7	7	8	8	8	6	7	9	7.5	7.5

Table 6 Final decision matrix

A^k	C $	RM (GB)	HD (TB)	PS (GHZ)	P	ASS
A^1	289.99	4	1	2.3	7.5	8.5
A^2	379.99	4	0.75	3	7.25	8.5
A^3	349.99	8	1	2.4	7.75	7.5
A^4	279.99	4	0.5	1.6	7.5	7.5
Optimization	Min	Max	Max	Max	Max	Max
A+	279.99	8	1	3	7.75	8.5
A-	379.99	4	0.5	1.6	7.25	7.5

4 GB of RAM, only 0.5 TB of hard disk capacity, a processor speed of 1.6 GHz, a prestige score of 7.25 and an after-sales service score of 7.5 on a scale from 1 to 9.

3.3.4 Attributes Weighting—Using AHP

As mentioned in the methodology section, TOPSIS finds the minimum weighted distance between the alternatives under evaluation and an ideal alternative. However, it also seeks to maximize the weighted distance to a non-ideal solution. This weighted distance implies that the attributes have a different level of importance. Therefore, AHP technique is used to determine such level. That said, in this case study each decision group member creates an array of pairwise comparisons as shown in Eq. (1) according to the scale illustrated in Table 2. These matrices are merged in one by using Eq. (4), which is evaluated using Expert Choice software, version 11.5, according to Fig. 3.

Table 7 illustrates the importance levels obtained in each attribute. According to such values, the family members give significant importance to the processor's speed, since its value is the highest. On the other hand, due to the high reliability of current computer systems, after-sales service is ranked as the least important and thus shows the lowest weight.

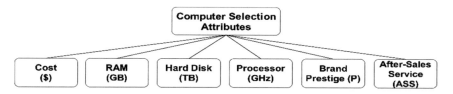

Fig. 3 Attributes and their level of importance

Table 7 Importance level of attributes

Attribute	C	RM	HD	PS	P	ASS
W	0.156	0.190	0.169	0.242	0.146	0.097

3.3.5 Normalization of Attributes

Since the attributes are expressed in different units of measure, we need to nondimensionalize these values. To this end, we employ the Euclidean norm of each attribute, considering that these are points in the k-dimensional space (i.e. there are k alternatives under evaluation) as illustrated in Eq. (6). This norm is calculated using Eq. (10). Table 8 illustrates the process.

The norm shown in the last line of Table 8 is used to nondimensionalize all the attribute values according to Eqs. (11)–(13), respectively. Table 9 shows the dimensionless attributes so they can be used in further operations.

3.3.6 Weighting Normalized Attributes

Before estimating the distances between the alternatives, it is convenient to weigh the attributes according to their importance levels, which were introduced in Table 7. In that sense, Table 10 illustrates the results obtained from this process.

3.3.7 Calculating Distances to the Ideal Solution

Once the normalized and weighted values are obtained, the next step is to estimate the distance of each alternative to the ideal solution and to the non-ideal alternative. Table 11 shows the calculated distances between every alternative and the ideal solution. The calculations were performed by using Eq. (14). The first column of the table includes the squared difference of each of the points, the sum, and the square root of the sum of those squares. The last column shows R, representing the rank of each alternative.

Note that at least one attribute in each alternative has no value, which indicates that the value of the attribute in that alternative is the ideal alternative. Likewise, based on the information presented in the table, it can be concluded that alternative A^3 is the best option and must be selected. In fact, A^3 has three attributes whose values fall into the ideal alternative value. On the other hand, A^4 seems to be the worst alternative to choose if we consider its distance to the ideal alternative.

Table 12 illustrates the distance values of each alternative with respect to the non-ideal alternative. Such distances were calculated using Eq. (15). Note that the preference order in column R is consistent. Thus, based on the distance values, alternative A^3 must be selected.

Table 8 Euclidean norm of attributes

A^k	C $	RM (GB)	HD (TB)	PS (GHZ)	P	ASS
A^1	84,094.2001	16	1	5.29	56.25	72.25
A^2	144,392.4001	16	0.5625	9	52.5625	72.25
A^3	122,493.0001	64	1	5.76	60.0625	56.25
A^4	78,394.4001	16	0.25	2.56	56.25	56.25
Sum	429,374.0004	112	2.8125	22.61	225.125	257
Norm	655.2663584	10.58300524	1.677050983	4.754997371	15.00416609	16.03122

Table 9 Normalization of attributes

A^k	C $	RM (GB)	HD (TB)	PS (GHZ)	P	ASS
A^1	0.44255	0.37796	0.59628	0.48370	0.49986	0.53022
A^2	0.57990	0.37796	0.44721	0.63092	0.48320	0.53022
A^3	0.53412	0.75593	0.59628	0.50473	0.51652	0.46784
A^4	0.42729	0.37796	0.29814	0.33649	0.49986	0.46784
A^+	0.42729	0.75593	0.59628	0.63092	0.51652	0.53022
A^-	0.57990	0.37796	0.29814	0.33649	0.48320	0.46784

Table 10 Normalized and weighted attributes

A^k	C $	RM (GB)	HD (TB)	PS (GHZ)	P	ASS
A^1	0.06904	0.07181	0.10077	0.11706	0.07298	0.05143
A^2	0.09046	0.07181	0.07558	0.15268	0.07055	0.05143
A^3	0.08332	0.14363	0.10077	0.12215	0.07541	0.04538
A^4	0.06666	0.07181	0.05039	0.08143	0.07298	0.04538
A^+	0.06666	0.14363	0.10077	0.15268	0.07541	0.05143
A^-	0.09046	0.07181	0.05039	0.08143	0.07055	0.04538

Table 11 Distance to the ideal alternative

A^k	C	RM	HD	PS	P	ASS	Sum	Dist	R
A^1	0.00001	0.00516	–	0.00127	0.00001	–	0.00644	0.08024	3
A^2	0.00057	0.00516	0.00063	–	0.00002	–	0.00638	0.07989	2
A^3	0.00028	–	–	0.00093	–	0.00004	0.00125	0.03531	1
A^4	–	0.00516	0.00254	0.00508	0.00001	0.00004	0.01282	0.11320	4

Table 12 Distance to the non-ideal alternative

A^k	C	RM	HD	PS	P	ASS	Sum	Dist	R
A^1	0.00046	0.00000	0.00254	0.00127	0.00001	0.00004	0.00431	0.06565	3
A^2	0.00000	0.00000	0.00063	0.00508	0.00000	0.00004	0.00575	0.07582	2
A^3	0.00005	0.00516	0.00254	0.00166	0.00002	0.00000	0.00943	0.09710	1
A^4	0.00057	0.00000	0.00000	0.00000	0.00001	0.00000	0.00057	0.02393	4

3.3.8 Calculating the Decision Index

Calculating the decision index in this case is not significant, since the order of preference defined by each alternative's distance to the ideal alternative and the non-ideal alternative is the same. However, for practical purposes, Table 13 shows the decision index of each alternative calculated using Eq. (16). Again, the best alternative to be selected is A^3.

Table 13 Decision index for the evaluated alternatives

A^k	Dist A^+	Dist A^-	Index	Order
A^1	0.08024	0.06565	0.55000	3
A^2	0.07989	0.07582	0.51308	2
A^3	0.03531	0.09710	0.26667	1
A^4	0.11320	0.02393	0.82549	4

4 Conclusion

Since the moment [44] proposed the first computer evaluation and selection methodology, it was already considered a complex problem. Today, after 40 years, selecting the computer equipment that best meets our particular needs is still challenging if we take into account the variety of features that modern systems include [45]. That said, current limitations in and issues with computer selection are not exclusive to a specific sector or domain. As, [46] points out, these types of problems are faced by many medical domains, since the knowledge of medical computer systems is lower than the knowledge of engineering systems.

This research proposes an AHP- and TOPSIS-based computer selection framework. On one hand, AHP is used to weight the attributes of the alternatives; on the other hand, TOPSIS helps the framework to select the best of such alternatives with respect to an ideal alternative. To exemplify the feasibility and practicality of our approach, we present a case study of a four-member family that wants to purchase a desktop computer. The major findings of our work can be highlighted as follows:

- Our AHP-TOPSIS-based methodology is easy and quick, since it is based on the intuitive concepts of ideal and non-ideal alternatives.
- The computer evaluation is performed by a decision group, which encourages a more democratic and inclusive decision-making environment.
- The decision group is formed by the family members, who best know their own needs with respect to using and purchasing a computer.
- Our AHP-TOPSIS-based methodology effectively integrates quantitative and qualitative attributes and thus gets rid of the shortcomings of traditional economic techniques that focus only on tangible attributes.

Although choosing "the perfect computer" seems rather straightforward, it may become a tedious process if it is not well conducted. As observed in the case study, reaching a consensus in the decision group is key to selecting the most important requirements of a computer. At the end of the research, the family purchased the best alternative, known as A^3, since its attributes reported the largest benefits. This demonstrates that the AHP-TOPSIS-based methodology that we propose successfully supports computer selection and can be adopted in other decision-making contexts, including supplier and human resources evaluation and selection.

5 Future Research

This work reports the theoretical framework for desktop computer selection and presents a case of study to illustrate every stage and activity performed. For future research, we will seek to use other multi-attribute techniques to improve the computer selection procedure or framework. Also, we will seek to integrate that techniques in a free distribution software for non-commercial uses.

References

1. Cingi, C.C.: Computer aided education. Procedia—Soc. Behav. Sci. **103**, 220–229 (2013)
2. Yigit, T., Koyun, A., Yuksel, A.S., Cankaya, I.A.: Evaluation of blended learning approach in computer engineering education. Procedia—Soc. Behav. Sci. **141**, 807–812 (2014)
3. Simkova, M.: Using of computer games in supporting education. Procedia—Soc. Behav. Sci. **141**, 1224–1227 (2014)
4. Patterson, R.W., Patterson, R.M.: Computers and productivity: evidence from laptop use in the college classroom. Econ. Educ. Rev. **57**, 66–79 (2017)
5. Challenges, C.N.: 2016 BSA Global Cloud. (2016)
6. Gartner: Quarterly personal computer (PC) vendor shipments worldwide, from 2009 to 2017, by vendor (in million units). Statista—the statistics portal. https://www.statista.com/statistics/263393/global-pc-shipments-since-1st-quarter-2009-by-vendor/
7. Chen, S.-M., Huang, Z.-C.: Multiattribute decision making based on interval-valued intuitionistic fuzzy values and particle swarm optimization techniques. Inf. Sci. (Ny) **397–398**, 206–218 (2017)
8. Chen, S.-M., Cheng, S.-H., Chiou, C.-H.: Fuzzy multiattribute group decision making based on intuitionistic fuzzy sets and evidential reasoning methodology. Inf. Fusion. **27**, 215–227 (2016)
9. Ekel, P., Kokshenev, I., Parreiras, R., Pedrycz, W., Pereira, J.: Multiobjective and multiattribute decision making in a fuzzy environment and their power engineering applications. Inf. Sci. (Ny) **361–362**, 100–119 (2016)
10. Ahn, B.S.: Approximate weighting method for multiattribute decision problems with imprecise parameters. Omega (United Kingdom) **72**, 87–95 (2017)
11. Taha, R.A., Choi, B.C., Chuengparsitporn, P., Cutar, A., Gu, Q., Phan, K.: Application of hierarchical decision modeling for selection of laptop. In: PICMET '07–2007 Portland International Conference on Management of Engineering & Technology, pp. 1160–1175. IEEE (2007)
12. García, J.L., Alvarado, A., Blanco, J., Jiménez, E., Maldonado, A.A., Cortés, G.: Multi-attribute evaluation and selection of sites for agricultural product warehouses based on an analytic hierarchy process. Comput. Electron. Agric. **100**, 60–69 (2014)
13. Chuu, S.-J.: Selecting the advanced manufacturing technology using fuzzy multiple attributes group decision making with multiple fuzzy information. Comput. Ind. Eng. **57**, 1033–1042 (2009)
14. Ertuğrul, İ., Karakaşoğlu, N.: Comparison of fuzzy AHP and fuzzy TOPSIS methods for facility location selection. Int. J. Adv. Manuf. Technol. **39**, 783–795 (2008)
15. Singh, H., Kumar, R.: Hybrid methodology for measuring the utilization of advanced manufacturing technologies using AHP and TOPSIS. Benchmarking An Int. J. **20**, 169–185 (2013)

16. Kreng, V.B., Wu, C.-Y., Wang, I.C.: Strategic justification of advanced manufacturing technology using an extended AHP model. Int. J. Adv. Manuf. Technol. **52**, 1103–1113 (2011)
17. Anvari, A., Zulkifli, N., Sorooshian, S., Boyerhassani, O.: An integrated design methodology based on the use of group AHP-DEA approach for measuring lean tools efficiency with undesirable output. Int. J. Adv. Manuf. Technol. **70**, 2169–2186 (2014)
18. Bai, C., Dhavale, D., Sarkis, J.: Integrating Fuzzy C-Means and TOPSIS for performance evaluation: an application and comparative analysis. Expert Syst. Appl. **41**, 4186–4196 (2014)
19. Lima-Junior, F.R., Carpinetti, L.C.R.: Combining SCOR®model and fuzzy TOPSIS for supplier evaluation and management. Int. J. Prod. Econ. **174**, 128–141 (2016)
20. Maldonado-Macías, A., Alvarado, A., García, J.L., Balderrama, C.O.: Intuitionistic fuzzy TOPSIS for ergonomic compatibility evaluation of advanced manufacturing technology. Int. J. Adv. Manuf. Technol. **70**, 2283–2292 (2014)
21. Zyoud, S.H., Fuchs-Hanusch, D.: A bibliometric-based survey on AHP and TOPSIS techniques. http://www.sciencedirect.com/science/article/pii/S0957417417300982?via%3Dihub (2017)
22. Pelorus, Karahalios, H.: The application of the AHP-TOPSIS for evaluating ballast water treatment systems by ship operators. Transp. Res. Part D Transp. Environ. **52**, 172–184 (2017)
23. Kusumawardani, R.P., Agintiara, M.: Application of Fuzzy AHP-TOPSIS method for decision making in human resource manager selection process. In: Procedia Comput. Sci., 638–646. Elsevier (2015)
24. Sindhu, S., Nehra, V., Luthra, S.: Investigation of feasibility study of solar farms deployment using hybrid AHP-TOPSIS analysis: case study of India. http://www.sciencedirect.com/science/article/pii/S1364032117301405?via%3Dihub (2017)
25. Carmone, F.J., Kara, A., Zanakis, S.H.: A Monte Carlo investigation of incomplete pairwise comparison matrices in AHP. Eur. J. Oper. Res. **102**, 538–553 (1997)
26. Gass, S.I., Rapcsák, T.: Singular value decomposition in AHP. Eur. J. Oper. Res. **154**, 573–584 (2004)
27. Lai, S.K.: A preference-based interpretation of AHP. Omega **23**, 453–462 (1995)
28. Russo, R.D.F.S.M., Camanho, R.: Criteria in AHP: a systematic review of literature. Procedia Comput. Sci., 1123–1132. Elsevier (2015)
29. Dymova, L., Sevastjanov, P., Tikhonenko, A.: A direct interval extension of TOPSIS method. Expert Syst. Appl. **40**, 4841–4847 (2013)
30. Nădăban, S., Dzitac, S., Dzitac, I.: Fuzzy TOPSIS: a general view. Procedia Comput. Sci., 823–831. Elsevier (2016)
31. Zhou, S., Liu, W., Chang, W.: An improved TOPSIS with weighted hesitant vague information. Chaos, Solitons Fractals **89**, 47–53 (2015)
32. Liang, D., Xu, Z.: The new extension of TOPSIS method for multiple criteria decision making with hesitant Pythagorean fuzzy sets. Appl. Soft Comput. **60**, 167–179 (2017)
33. Saaty, T.L.: Time dependent decision-making; dynamic priorities in the AHP/ANP: Generalizing from points to functions and from real to complex variables. Math. Comput. Model. **46**, 860–891 (2007)
34. Saaty, T.L.: Decision-making with the AHP: why is the principal eigenvector necessary. Eur. J. Oper. Res. **145**, 85–91 (2003)
35. Saaty, T.L.: Highlights and critical points in the theory and application of the analytic hierarchy process. Eur. J. Oper. Res. **74**, 426–447 (1994)
36. Escobar, M.T., Aguarón, J., Moreno-Jiménez, J.M.: A note on AHP group consistency for the row geometric mean priorization procedure. Eur J Oper Res. **153**(2), 318–322 (2004)
37. Dong, Y., Zhang, G., Hong, W.C., Xu, Y.: Consensus models for AHP group decision making under row geometric mean prioritization method. Decis. Support Syst. **49**, 281–289 (2010)

38. Chen, M.F., Tzeng, G.H.: Combining grey relation and TOPSIS concepts for selecting an expatriate host country. http://www.sciencedirect.com/science/article/pii/S0895717707050000 75?via%3Dihub (2004)
39. Rudnik, K., Kacprzak, D.: Fuzzy TOPSIS method with ordered fuzzy numbers for flow control in a manufacturing system. Appl. Soft Comput. J. **52**, 1020–1041 (2017)
40. Lee, K.-K., Lee, K.-H., Woo, E.-T., Han, S.-H.: Optimization process for concept design of tactical missiles by using pareto front and TOPSIS. Int. J. Precis. Eng. Manuf. **15**, 1371–1376 (2014)
41. Mao, N., Song, M., Deng, S.: Application of TOPSIS method in evaluating the effects of supply vane angle of a task/ambient air conditioning system on energy utilization and thermal comfort. Appl. Energy **180**, 536–545 (2016)
42. Li, X., Chen, X.: Extension of the TOPSIS method based on prospect theory and trapezoidal intuitionistic fuzzy numbers for group decision making. J. Syst. Sci. Syst. Eng. **23**, 231–247 (2014)
43. Çevik Onar, S., Büyüközkan, G., Öztayşi, B., Kahraman, C.: A new hesitant fuzzy QFD approach: an application to computer workstation selection. Appl. Soft Comput. J. **46**, 1–16 (2016)
44. Timmreck, E.M.: Computer selection methodology. ACM Comput. Surv. **5**, 200–222 (1973)
45. Poynter, D.: Computer selection guide (1983)
46. Preston, J.D.: Guide to computer selection (1994)

An Agent-Based Memetic Algorithm for Solving Three-Level Freight Distribution Problems

Conrado Augusto Serna-Urán, Martín Darío Arango-Serna,
Julián Andrés Zapata-Cortés and Cristian Giovanny Gómez-Marín

Abstract Vehicle routing problems for the transport of people or goods are combinatorial problems belonging to the nondeterministic polynomial time (NP-Hard) category. This means that the solution is very limited if using mathematical programming or combinatorial optimization procedures. As alternative solutions, several metaheuristics have been developed, including memetic algorithms that explore possible solution spaces by implementing efficient procedures for vehicles routing in real applications. However, the combinatorial problem is still present and these solution techniques lose efficiency if the search space is enlarged or if shorter solution times are needed—both common demands in real-world applications. This chapter presents an evolutionary memetic type metaheuristic in a multi-agent system that combines global and local search strategies for solving a three-level freight distribution network. The proposed model uses coordination and collaboration strategies between several agents, to improve the performance of the freight transport process and use a specific gender alternative at the memetic algorithm that reinforces the metaheuristic evolutionary process. The results obtained by multi-agent system model application are compared with the Solomon insertion heuristic, generating better solutions that improve the distribution process in terms of less total travel distance and variability.

Keywords Memetic algorithm · Multi-agent system · Vehicle routing problem

C. A. Serna-Urán
Universidad de San Buenaventura, Medellín, Antioquia, Colombia
e-mail: conrado.serna@usbmed.edu.co

M. D. Arango-Serna · C. G. Gómez-Marín
Universidad Nacional de Colombia—Sede Medellin, Medellín, Antioquia, Colombia
e-mail: mdarango@unal.edu.co

C. G. Gómez-Marín
e-mail: crggomezma@unal.edu.co

J. A. Zapata-Cortés (✉)
Institución Universitaria CEIPA, Sabaneta, Antioquia, Colombia
e-mail: Julian.zapata@ceipa.edu.co

© Springer International Publishing AG 2018
R. Valencia-García et al. (eds.), *Exploring Intelligent Decision Support Systems*,
Studies in Computational Intelligence 764,
https://doi.org/10.1007/978-3-319-74002-7_6

111

1 Introduction

The computer processing capabilities developed in the last decades enabled shorter computational times, thereby facilitating the solution to complex mathematical problems such as those that model logistic distribution processes. This development, combined with the emerging of solution techniques such as metaheuristics, has positively impacted logistics operations, making it more efficient. However, better metaheuristics and higher capacity computer systems fall short when it comes to modeling complete distribution process dynamics where multiple logistics actors are to be integrated. Adding to the difficulty, and having a considerable effect on logistics planning and control, is the increasing urban complexity in freight transport, the variety of transport modes and the permanent changes in customers' needs [1].

This creates the need to develop new methodologies allowing models to include these contexts, improving logistics efficiency and flexibility as well as understanding the capabilities, goals and needs of each stakeholder involved in the logistics processes [2]. In this context, multi-agent systems have emerged as a very promising technique that facilitates different multiparty coordination in very complex scenarios. This system can be integrated with mathematical procedures in order to enhance the model solutions [3–5].

Multi-agent systems are a new paradigm of distributed computing that can be used to represent a large number of real world situations, including problems related to goods distribution. In Multi-agent systems, different actors involved in logistics operations can be considered as agents integrated and coordinated in order to achieve common objectives based on their particular goals and capabilities [6, 7]. Multi-agent systems generate parallel and coordinated solutions, which provide dynamic solutions to complex problems, difficult to solve using traditional methods [8].

A supply chain is made of several actors that interact and present different dynamic behaviors. Decision-making in this context involves handling a large information amount from different actors. It is necessary to include related processes that are frequently slow and poorly coordinated. This can be overcome by considering entities that interact smartly between them. These entities can provide domain-specific information to validate the decision-making system at tactical and operational levels. Multi-agent systems approaches are very good for this [9]. In this sense, logistics network can be understood as a set of intelligent collaborative agents [10], interacting with each other and responsible for one or more activities [11].

Agents can be related to the planning or execution logistics processes to improve decision making in dynamic and complex environments [12]. On the other hand, multi-agent systems are a useful methodology for evaluating the multi-objective nature of logistics systems, and for the study of the behavior of several stakeholders affected by the distribution/logistics processes and environmental changes [13, 14].

In this chapter, a Multi-Agent model is presented aiming to solve a three-level logistic distribution problem, which involves a network made of several large-scale hubs and smaller distribution centers. In order to increase the multi-agent system efficiency, a solution method based on a memetic algorithm is integrated with the exploration of local and global search spaces. The result is a multi-agent system for solving a three-level goods distribution problem with time windows.

2 The Three-Level Logistic Distribution Problem

The urban distribution of goods is a complex process that includes all freight transport with its associated operations such as delivering, collection, cargo transfers, loading/unloading, points of sale placing, cross docking, and returns in reverse logistics [15]. This process also requires a consideration of restrictions related to infrastructure and other cities [12]. To overcome these difficulties, there are several distribution strategies that companies can use to improve their distribution processes [13, 16, 17].

One of the most used and studied strategy is the setting of distribution centers, in which goods from several suppliers are consolidated and then delivered to customers. As the complexity of the distribution process increases, e.g. due to the presence of a large customer base or due to the configuration of a clusters set of them, there arises the need for more consolidation centers in order to optimize the transport process. This is known a multi echelon distribution process [18–20].

Freight companies operating in urban areas usually structure their distribution network with hubs (Large scale distribution centers) in which goods from suppliers are consolidated and then sent to a set distribution centers close to residential areas. The cargo is divided into small quantities in these hubs, and subsequently delivered to the end customers. This logistic structure, as depicted in Fig. 1, reduces the cost of transportation by enabling the optimization of distances for product delivery from suppliers to customers [19].

This logistics network involves two levels for cargo consolidation and a third level in which the goods are sent to the customers. In each level the transport process must be optimized. This network is known as a Three-Level Logistic Distribution Problem, which has to be globally optimized using complex enough mathematical formulations and solving procedures [21].

Fig. 1 Logistics network. *Source* [19]

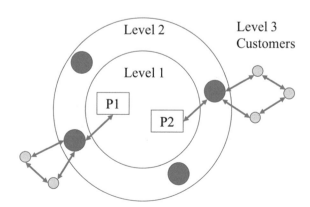

2.1 Mathematical Formulation Three-Level Logistic Distribution Problem

Laporte [19] modeled the Three-Level Logistic Distribution Problem and presented a mathematical formulation, later used and updated by [22]. In order to optimize the distribution costs for the three level problem transport activities, it is necessary to consider the following optimization objectives:

- Selecting locations for facilities on the second level.
- Establishing a vehicles fleet based upon the facilities.
- Building the optimal delivery routes from the first level to the second and from it to the customers.

The nomenclature of the parameters and variables require for the Three-Level Logistic Distribution model is presented in Table 1 [22].

As seen from the decision variables, the model has to locate the facilities, assign the vehicles and accommodate the routes that optimize the distribution process costs. The problem can then be formulated as:

Table 1 Three-level logistic distribution model parameters and variables

Model parameters	
ζ_{ij}	Distance between node i and j
β	Vehicle capacity
c_{ij}	Travel cost from j to j
s_i	Service length
τ_i	Upper time limit to serve i
e_i	Lower time limit to serve i
d_i	Demand/supply at i
t_{ij}	Travel time to i and j
g_h	Unit handling cost in facility h
V_h	Maximum capacity of facility h
τ_j	Vehicle unload time in j
E	Maximum distance for the vehicle
T	Maximum operation time of the vehicle
θ	Unit distance cost of the vehicle
η_{ij}	Fixed cost of delivering products from i to j
Decision variables	
x_{ij}	Binary variable equal to 1 if the vehicle travel from h to n; 0 otherwise
z_{hn}	Binary variable equal to 1 if customer n is served from the hub h; 0 otherwise
a_{fh}	Integer variable. Amount of goods to be sent from f to facility h

Source Own source

$$\min \sum_{f \in F} \sum_{h \in H} c_{fh} a_{fh} + \sum_{h \in H} \sum_{n \in N} (g_h d_n + \eta_{hn}) z_{hn}$$
$$+ \sum_{i \in H \cup N} \sum_{j \in H \cup N} \theta \zeta_{ij} x_{ij} \tag{1}$$

Restricted to:

$$\sum_{i \in H \cup N} x_{ij} = 1 \quad \forall j \in N \tag{2}$$

$$\sum_j \sum_{i \in H \cup N} d_j x_{ij} \leq \beta \tag{3}$$

$$\sum_{i \in H \cup N} \sum_{j \in H \cup N} c_{ij} x_{ij} \leq E \tag{4}$$

$$\sum_{i \in H \cup N} \sum_{j \in H \cup N} \tau_j x_{ij} + \sum_{i \in H \cup N} \sum_{j \in H \cup N} t_{ijk} x_{ij} \leq T \tag{5}$$

$$\sum_{i \in S} \sum_{j \in (H \cup N) - S} x_{ij} \geq 1 \tag{6}$$

where $2 \leq |S| \leq |H \cup N|; \quad S \subseteq |H \cup N|; S \cap N \neq 0$

$$\sum_{i \in H} \sum_{j \in N} x_{ij} \leq 1 \tag{7}$$

$$\sum_{j \in (H \cup N)} x_{ij} - \sum_{j \in (H \cup N)} x_{ji} = 0 \quad \forall i \in (H \cup N) \tag{8}$$

$$\sum_{f \in F} a_{fi} - \sum_{j \in N} d_j z_{ij} = 0 \forall i \in H \tag{9}$$

$$\sum_{f \in F} a_{fi} \leq V_i y_i \quad \forall i \in H \tag{10}$$

$$-z_{ij} + \sum_{u \in H \cup N} x_{iu} + x_{uj} \leq 1 \quad \forall i \in H; \quad j \in N \tag{11}$$

$$a_{fi} \geq 0 \tag{12}$$

The objective function (1) aims to minimize the cost of freight transport to the hubs, the cost of handling the freight and the cost of travel distance in the final delivery tour. The constraint number (2) ensures that all customers are visited only once. Equations (3–5) are related with the operation of the vehicles such as capacity, operation time and maximum distance. Restriction (6) ensures that at least one vehicle leaves the terminals, while restriction (7) limits visits to the node only from the hub i. Constraint (8) is a flow balance between the inputs and outputs of node i. Equation (9) ensures the relationships of the load and demand transported between suppliers, hubs, and customers. Constraint 10 guarantees that the amount transported from suppliers to hubs be equal or less than the hub i capacity. Equation (11) is an equilibrium constraint between the trips that start and finish at hub i. Constraint 12 ensures that the carrying capacity of fi is greater or equal than zero.

This model is part of the Location-Routing problem, which can be solved using exact and heuristic techniques, in many cases with relative successes. In the case of exact methods, the difficulty arises that these problems are NP-complete, in which finding an optimal solution involves evaluating search spaces not reachable with current techniques and computing systems [23]. On the other hand, the complexity of the real-life data makes the solution process even more difficult. In these situations, it is possible to apply heuristic approximation algorithms that produce feasible results, but do not guaranty finding the optimal solutions for all cases [24].

The mathematical model for the Three-Level Logistic Distribution Problem, presented above, is NP-hard, meaning that it is very difficult to solve and that requires the use of more robust techniques. Memetic algorithms have proven to be very effective in solving complex problems, but this may be insufficient due to flexibility needs in dynamic contexts [25, 26]. In addition, the individual decisions of each stakeholder in the goods distribution process must be analyzed carefully, because it may affect the whole system.

Transport systems can contain thousands of autonomous and intelligent entities that must be simulated and/or controlled [27–29]. Therefore, it is desirable to use techniques that facilitate coordination between distribution processes actors. Through this it will be possible to generate more efficient transport plans than those generated individually [22]. Multi-agent structures are a flexible methodology for solving logistics problems that include several stakeholders and entities. They are becoming an interesting alternative to model this kind of real processes [30, 31].

2.2 Multi-Agent Systems for Solving Logistics Problems

The Multi-Agent System (MAS) is a system composed of multiple interacting intelligent agents, which are autonomous entities with the ability to distinguish, perceive and take action while incorporating the interactions with other agents [14]. MAS is a useful methodology that allows considering several objectives involved in logistics and transportation due to the amount of actors involved in such processes [14]. MAS are especially good for large scale and dynamic problems due to their ability to make quick local decisions [32].

In the scientific literature, it is possible to find many works that designed multi-agent systems to represent logistic processes. Schroeder et al. [33] present a multi-agent model that represents logistics decisions such as freight consolidation, distribution and pick-up and delivery, in an integrated commercial and passenger transport system.

Maciejewski and Nagel [34] model a dynamic vehicle routing problem using a platform that allows developing microsimulation models based on computer agents called Matsim. Barbucha [35] solve a capacitated vehicle routing problem in distributed and undistributed environments. In its work, [4] optimize a vehicle routing problem with limited capacity in a two-level distribution problem through multi-agent modeling using Constrain Programing (CP) and Mixed Integer Programing (MIP).

Wangapisit et al. [14] modeled an urban distribution center and the administration of the cargo vehicle parking through a multi-agent model with learning reinforcement, generating city logistics policies that can reduce the operation cost and minimize environmental impact. A self-adaptive multi-agent architecture for outsourcing operations in midsize companies, which mitigates the effects of uncertainty in supply chains was designed by [36]. van Lon and Holvoet [32] use multi-agent system to solve the dynamic pick-up and delivery vehicle routing problem with time windows with different degrees of dynamism, urgency and scale.

Matteis et al. [37] modeled the interactions between freight generators, shippers and freight receivers, which allow to explain the delivery structures in urban freight transport. Arango-Serna and Serna-Uran [38] formulated a new fuzzy-based negotiation protocol for assigning service orders in a freight distribution network.

Anand et al. [6] established the methodological relationships between the urban logistics characteristics and multi-agent modeling. They also designed the stages for the successful implementation of this type of modeling so that the interactions between the actors and the distributed decision-making process dynamics could be captured. Ghadimi et al. [39] developed a multi-agent model for supplier selection and order allocation in an integrated supply chain. Through a multi-agent model, [40], evaluated the vulnerability of the multiple urban goods distribution actor's objectives of the unexpected situations such as floods or rain.

Those works are focused on solving two-Level distribution networks. However, transport networks usually involves at least one distribution/CrossDock center in order to improve the logistics network performance, making it necessary to include at least another decision level with one or several distribution/CrossDock facilities. For that reason, the approaches proposed by the abovementioned authors are limited to simple distribution networks. With the aim to propose a more general and realistic model, an Agent-based memetic algorithm is proposed in the next sections, which not only models the three-level logistic network including several suppliers, distribution/CrossDock facilities and customers, but also presents a new coordination and cooperation processes between agents. The combination of memetic algorithms with multi-agent systems is a novel approach for solving this type of distribution problems.

3 Methodology

The Three-Level Logistic Distribution Problem is solved in this chapter using an Agent-Based Memetic Algorithm, with the aim of avoiding the abovementioned exact and heuristic techniques restrictions. This Agent-Based Memetic Algorithm also allows considering city logistics conditions, which as in real life, require the interaction between several distribution system actors.

The network used as the test scenario is composed of 240 customers, which are served by 28 terminals grouped in 4 hubs. The multi-agent model and the agent based memetic algorithm was performed on the Java Agent Development Framework (JADE)© using Eclipse© java interface. A total of 25 runs were performed and the memetic algorithm made 5000 iterations at each run. The multi-agent Model is presented in this section, highlighting the agent relationships. The proposed memetic algorithm is presented below.

3.1 Multi-agent Model

Four types of agents compose the multi-agent model. This model configures a communication architecture and integration process which are performed in a decentralized and coordinated operation framework. Figure 2 depicts a model representation. It starts with a customer that makes a request and activates a protocol set in the internal agents that are organized according to its nature. These agents are the control agent, infrastructure agents, hub agents and vehicle agents.

Each agent has a set of behaviors as defined below. These behaviors configure a modular and flexible system that can be adjusted to different conditions according to the agents' goals and capacities.

The description of each agent used in the model is:

- **Agent Customers**: These agents are the initiators of the logistic distribution pro-cess. The demand needed to be satisfied for every customer becomes a service order that includes basic information for the subsequent planning

Fig. 2 Proposed multi-agent model. *Source* Own source

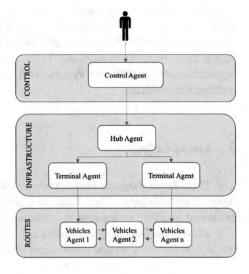

process, such as: products type and quantity required, customer location and time windows to be served.

- **Agent Control**: The control agent accepts or rejects the customers' requests according to the available capacity and it also assign priorities to these orders. Another function is to establish service areas according to the projected demand and the available logistics infrastructure.
- **Agent Infrastructure**: This agent is in charge of assigning and coordinating the movement of products from the hubs and terminals to the customers. The Hubs Agents and the Terminals Agents make up this category.
- **Agent Vehicle**: The Agent vehicle is responsible of defining and optimizing the required routes for the products movement from their origin to their destination through the entire logistics network. In other words, it is in charge of defining the routing in every level of the distribution network.

Communication between agents is a key process in the coordination of multi-agent processes. Coordination in the multi-agent system facilitates the support process that ensures that the individual actions of each agent contribute to the

Table 2 Model of behaviors according to protocols FIPA

Agent	Behavior name	Behavior type	Description of the task
Agent control	Enter Customer( )	SimpleBehaviour( )	Allows the agent to welcome the customer and accept their request, in order to initiate delivery operation through the hub
	Acrossover( )	CyclicBehaviour/ TickerBehaviour	Coordinates the execution of evolution operators. It makes the call to synchronously activate the communication protocols between the agents that will execute the selection and crossover processes
Agent hubs	Asign( )	WakerBehaviour	This behavior allows the Hub agent to wait for a period of time. After that, it assigns the orders to each terminal according to its location and capacity
Agent terminal	Consolidate( )	WakerBehaviour	This behavior allows the agent to wait for a period of time and to be assigned with the respective commands to initiate the distribution routes definition
	Solomon( )	OneShotBehaviour	It begins the routes definition process, using the Solomon heuristic. It also ensures the solutions feasibility according to the time windows accomplish
	NewAgentV( )	OneShotBehaviour	The routes originated in each terminal through the Solomon( ) behavior, are the basis to initiate the agents Vehicle creation (one agent for each route)
Agent vehicle	Memetic()	SequentialBehaviour	This behavior sequentially executes the sub-behaviors associated with the evolution operators

(continued)

Table 2 (continued)

Agent	Behavior name	Behavior type	Description of the task
	RecieveNode()	SimpleBehaviour	This behavior allows the agent to receive the information that updates the assigned orders through the evolution process
	Selection()	OneShotBehaviour	This behavior facilitates the selection process through the Contract Net communication protocols
	Crossover()	OneShotBehaviour	This behavior applies the corresponding crossover operator. It generates a N of number of descendant candidates
	Evaluation()	OneShotBehaviour	It evaluates the fitness value of each candidate and their feasibility according to the capacity and time windows constraints
	LocalSearch()	OneShotBehaviour	This behavior is in charge of executing the 2-opt heuristic
	Gender()	OneShotBehaviour	It randomly assigns the gender (female or male) to the feasible descendants and performs the registration process in the male agent's service directory

Source Own source

solution of the logistic model. In the proposed model, the communication between the agents is done using the FIPA-ACL standard and its interaction protocols [41]. Table 2 presents the behavior for every agent considered in the proposed Multi-agent Model.

The behaviors could be classified as being generic and compound. The Generic behaviors are: the SimpleBehavior that execute simple operations; the OneShotBehavior that represents behaviors that are executed just once; WakerBehavior is a sub class OneShotBehavior that allows that an agent waits to execute a task; CyclicBehavior that makes that an agent executes a task repeatedly. Compound behaviors are used to execute complex actions and determine sub-behaviors or child's behaviors [42]. The SequenceBehavior is used in the proposed multi-agent model, in which the child's behaviors are executed following simple sequences.

The communication strategy between the different levels of the network is made based on coordination protocols such as the Contract Net protocol (CNP) [43], which is a decentralized structure where the agents take bidders roles and contractors and also consider application protocols.

3.2 Agent-Based Memetic Algorithm

In a traditional genetic algorithm, the individual solutions do not evolve during their lifetime. They are created, evaluated, then possibly selected as new solutions parents and later destroyed. However, research on memetic and local genetic search algorithms has shown that performance can be improved if solutions evolve during their whole lifetime [44, 45].

Memetic algorithms (MA) belong to evolutionary algorithms (EA) that apply an independent local search process to improve the solutions. The combination of global and local search is a successful strategy in global optimization. In fact, MA has been recognized as a powerful evolutionary computing algorithm [46].

The relative advantage of MA over other EA is quite consistent in complex search spaces. The MA are inspired by adaptation models in natural systems that combine the evolutionary adaptation of populations with individual learning problems throughout life. In addition, memetic algorithms are inspired by the Dawkins' concept of a Meme, which represents a unit of the cultural evolution that can exhibit local refinement. From the point of view of optimization, memetic algorithms are hybrid evolutionary algorithms that combine global and local search using an AE to carry out exploration, while the local search method performs exploitation [47].

In the proposed model, each vehicle becomes an agent that evolves from the interaction with other vehicles of the same type. These vehicles are part of the hub to which they have been assigned. The multi-agent model follows the structure of agents proposed by the software Jade [42], in which the processes of communication and coordination from FIPA protocol are one of its most important elements. The evolution process use the following operators:

- Population start
  The Solomon() protocol is executed in each terminal to determine the routes that must to be followed by the vehicle agents. Vehicle agents are divided into an emitter population (vehicles m) and receiver population (vehicles f). Each population has genetic information for the crossover operation in order to develop a structured evolution process as explained in [48].
- Setting the evolution process
  The Control Agent initiates the crossover process by activating the Contract Net protocols on vehicles of type f to request offers between vehicles of type m in the same terminal. Offers are calculated randomly.
- Selection of offers
  Each vehicle of type f selects the best offer sent by the vehicles of type m that responded to the request.
- Crossover
  The crossover process is executed according to the Best Cost Route Crossover operator (BCRC) [22] with the selected route.

- Local Search
  A 2-opt algorithm develops the local search process. If the fitness value of the parent's routes is improved by the child, the evolutionary process continues, otherwise it will continue with the search. In the event that there is not improvement in the total travel cost, the child couple is eliminated and the original genetic information is back to the parent's agents.
- Route selection
  The resulting routes are evaluated for feasibility conditions in relation to the capacity and the compliance to the time windows. The feasible routes that decrease the distance traveled are distributed again among the vehicles involved in the crossing process.
- Mutation
  A mutation process is done in every crossover cycle. This is performed to facilitate the node exchange process between terminals. The mutation operator, defined as the randomized multi-terminal mutation operator (OMTA), has the following procedure.

Each terminal evaluates the distance of each assigned customer and assigns to it a terminal change value according to the function:

$$q_i = \sum_{i}^{N^t} p_i \quad i = 1, 2, 3, \ldots N^t \tag{13}$$

$$p_i = \frac{f(x_i)}{\sum_{1}^{N^t} f(x_i)} \tag{14}$$

where $f(x_i)$ is the distance from node i to the terminal and N^t is the number of nodes assigned to the terminal

- A client that meets the following condition is selected:

$$q_{i-1} < \varphi < q_i \tag{15}$$

where $\varphi = random()$.

- Randomly select a terminal between the other different terminals. Apply the insertion process in the routes that belong to the terminal, according to the insertion criterion iii described in [49].
- Select the best terminal by evaluating the joint traveled distance between the origin and the destination terminal.
- End the mutation process.
  The memetic evolution process followed by each route is described in the following code.

Code 6.1 Code for the Agent-based memetic algorithm. Source: Own Source

> *The Control Agent Executes the Acrossover() behavior*
> *For each Agent Terminal separate vehicles into two sets (m vehicles and f vehicles)*
> *Repeat until the end Process == True*
> *For each vehicle f, request a crossover offer to f vehicles at the same terminal*
> *The vehicle m evaluates the function Offer = random(), and sends the offer value.*
> *The vehicle f selects other vehicle m according to the function VehicleCrossover (minimum offer)*
> *Vehicle k executes the BCRC crossover operator BCRC → route 1 (r1), route 2 (r2)*
> *Execute 2-Opt Local Search for r1 and r1*
> *If capacity and time windows conditions are accomplished, and if function Fitness(ri, r2) < Fitness (vehicle f, vehicle m)*
> *Replace f by r1 and r2*
> *Send route r2 to vehicle f*
> *Execute the mutation operator*
> *End of process*

4 Results

As mentioned above, the proposed agent-based memetic algorithm is used to solve the three-level logistic distribution problem. The test scenario is composed of 240 customers, which are served by 28 terminals, which are grouped in 4 hubs. According to multi-agent model presented in the methodology section, the customers are assigned through the Contract Net protocol executed for the control agent. The model is executed 25 times and a comparison process is made between the solutions found with the Solomon insertion heuristic and the proposed agent-based memetic algorithm. The results obtained for the Solomon Heuristic to solve the scenario problem, as the initial routes to the memetic algorithm in one of the executions, are presented in Table 3.

Table 3 shows that the Solomon Heuristic generates 38 routes required to fulfill the customers' demand. Some terminals are served by the same hub as well as some routes starting in the same terminal, as in the case of routes 33, 34 and 35 that begins in terminal 26. The 38 routes require a total distance to be traveled of 4315 units.

The implementation process of the proposed agent-based memetic algorithm allows the hubs to exchange customers through the evolutionary process,

Table 3 Solomon solution for one execution

No	Hub	Terminal	Vehicle	Solomon routes	Distance
1	Hub 1	T1	VT1-0	[57 114 168 160 48 21 184]	130
2	Hub 1	T2	VT2-0	[104 224 44 96 149]	101
3	Hub 1	T3	VT3-1	[228]	22
4	Hub 1	T3	VT3-0	[19 117 209 16 200 223 180 22 1]	164
5	Hub 1	T4	VT4-0	[124 91 161 131 189 11 150 33 84 60 193]	187
6	Hub 1	T5	VT5-0	[90 51 129 4 211 39 194 15 154 75]	188
7	Hub 1	T5	VT5-1	[94 65 159 61]	84
8	Hub 2	T10	VT10-0	[103]	20
9	Hub 2	T11	VT11-0	[219 191 186 167 31]	100
10	Hub 2	T6	VT6-0	[6 222 8 74 195]	89
11	Hub 2	T7	VT7-0	[172 116 82 226]	74
12	Hub 2	T8	VT8-0	[230 28 83 181 135]	84
13	Hub 2	T9	VT9-0	[41 199 93 45 78 89 38]	142
14	Hub 3	T12	VT12-1	[158 151 101 35 203]	96
15	Hub 3	T12	VT12-0	[29 235 197 55 157 217 177 20 142 187 80 81]	194
16	Hub 3	T13	VT13-0	[183 148 227 79 188 115 214 105 231 59 102]	189
17	Hub 3	T13	VT13-1	[125 133]	37
18	Hub 3	T14	VT14-0	24 112 53 73 206 185 192]	120
19	Hub 3	T15	VT15-0	[12 9 190 32 170]	82
20	Hub 3	T16	VT16-0	[100 121 213 166 37]	70
21	Hub 3	T17	VT17-0	[141 72 54 221 238 56 153 165 210 173 36 143]	198
22	Hub 3	T17	VT17-1	[18 132]	22
23	Hub 3	T18	VT18-0	[85 225 64 106 162 30 207 202 212]	182
24	Hub 3	T19	VT19-0	[5 145 175 118 3 10 146 205 108]	185
25	Hub 3	T20	VT20-0	[88 92 86 58 220 155 164 70 176 50 109]	200
26	Hub 3	T20	VT20-1	[178 46]	43
27	Hub 4	T21	VT21-0	[23 17 119 63 196 67 163 99 71 144 26]	190
28	Hub 4	T22	VT22-1	[139]	18
29	Hub 4	T22	VT22-0	[137 62 174 198 136 182 113 77 49 87]	184
30	Hub 4	T23	VT23-1	[126 152]	41
31	Hub 4	T23	VT23-0		185

(continued)

Table 3 (continued)

No	Hub	Terminal	Vehicle	Solomon routes	Distance
				[236 156 171 13 42 138 140 111 107 120]	
32	Hub 4	T24	VT24-0	[169 25 234 97 69]	92
33	Hub 4	T25	VT25-1	[201 128 204 208 134 14]	125
34	Hub 4	T25	VT25-2	[218]	21
35	Hub 4	T25	VT25-0	[122 237 179 34 229 239 127 240 43 7 123 27]	198
36	Hub 4	T26	VT26-1	[110 98]	32
37	Hub 4	T26	VT26-0	[2 76 147 130 68 216 233 232 47 215 95]	193
38	Hub 4	T27	VT27-0	[52 40]	33
Total distance					4315

Source Own source

which facilitates the search spaces exploration. This allows exploring not only the search space of the terminal that contains the routes that is being optimized but also the other terminal spaces.

The communication process performed between the agents' vehicle in the memetic algorithm is depicted in Fig. 3. This figure shows one of the thousands of crossover events that occur in the model evolution process. In this case, agent vehicle VT25-1 sends a request to agent DF to initialize the crossover event. DF is one of the Jade's standard agents, used to coordinate the search of agents with willingness to initializes its crossover process. Agents VT25-0 and VT25-2 answers and send offers to VT25-1. As shown in the figure, agent VT25-2 is chosen to initialize the memetic evolution process.

The results obtained after implementing the agent-based memetic algorithm for solving the instance problem at the same execution of the Solomon Heuristics are presented in Table 4.

The results of the agent-based memetic algorithm presented in Table 4 show that the number of routes required in the proposed algorithm reduces the number of routes from 38 to 33. It can also be observed that the total distance is shortened from 4315 to 3579 units, which represents a 17% reduction. The decrease of routes number and total travel distance means a reduction in the vehicles use and in the costs directly associated to the distribution process.

As mentioned above, 25 runs (executions) were analyzed to validate the results. The solutions found were compared using the Solomon heuristic and the proposed agent-based memetic algorithm. The variability of the total distances required for the 25 executions is presented in Fig. 4 using a box-plot.

Figure 4 shows that the Solomon solutions have higher distances values for the 25 executions than the proposed agent-based memetic algorithm. It can also be

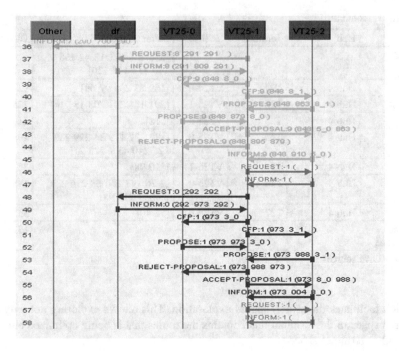

Fig. 3 Exchange of messages in routes evolution process

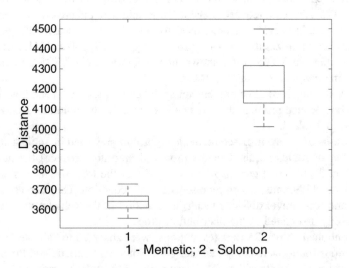

Fig. 4 Boxplot comparing Solomon Heuristic versus the proposed agent-based memetic algorithm solutions. *Source* Own source

Table 4 Proposed agent-based memetic algorithm solution

No	Terminal	Vehicle	Multi agent routes	Distance
1	T1	VT1-0	[114 57 168 160 48 21 184]	106
2	T10	VT10-0	[230 28 83 116 181]	75
3	T11	VT11-0	[219 191 186 167 31]	100
4	T12	VT12-0	[29 197 235 157 55 20 217 187 177 81 142 80]	143
5	T13	VT13-0	[183 148 133 125 79 105 227 214 188 231 59]	128
6	T14	VT14-0	[23 67 17 119 63 163 196 99 71 144 26]	157
7	T15	VT15-0	[5 145 175 118 3 10 205 146 108]	113
8	T16	VT16-0	[100 121 166 37 213 102]	83
9	T17	VT17-0	[12 9 190 56 115]	75
10	T18	VT18-0	[85 225 64 106 162 30 207 202 212]	182
11	T19	VT19-0	[141 54 72 221 238 153 165 173 210 36 132]	117
12	T2	VT2-0	[104 224 44 96 149]	81
13	T20	VT20-0	[88 92 86 58 155 220 164 109 176 70 50]	196
14	T20	VT20-1	[143 18 178 46]	65
15	T21	VT21-0	[66 24 112 53 192 185 73 206 32 170]	115
16	T22	VT22-0	[137 174 62 136 113 182 198 139 77 87]	118
17	T23	VT23-0	[236 156 171 13 138 140 111 120 107]	146
18	T23	VT23-1	[42 126 152]	59
19	T24	VT24-0	[169 25 234 97 69]	92
20	T25	VT25-1	[134]	18
21	T25	VT25-0	[122 237 179 229 239 43 34 128 240 27 7 123]	160
22	T25	VT25-2	[49 208]	36
23	T26	VT26-0	[76 147 130 2 47 68 216 232 233 110 215]	183
24	T26	VT26-1	[98 95]	34
25	T27	VT27-0	[201 52 40 127 218 204 14]	119
26	T3	VT3-0	[19 117 209 228 200 180 16 1 223 22]	180
27	T4	VT4-0	[124 91 161 131 189 11 150 33 193 84 60]	166
28	T5	VT5-0	[65 94 90 51 129 211 4 61 154 75]	120
29	T5	VT5-1	[39 194 159 15]	70
30	T6	VT6-0	[6 222 103 74 8 195 101]	106
31	T7	VT7-0	[158 203 35 151]	58
32	T8	VT8-0	[172 82 226 135]	63
33	T9	VT9-0	[41 199 93 45 78 89 38]	115
Total distance				3579

Source Own source

observed that the proposed model has less variation (lower dispersion) than the Solomon solutions. Lower variation is desirable as it assures a better behavior of the algorithm, in that it is more stable and precise.

The reduction of total distance and the variation by the proposed model in all the runs, can be explained by two reasons. The first is associated to the agent's behaviors and the communication and collaboration processes, which allow a better assignment of the cargo, vehicles and facilities, as a result of the interaction of all entities of the model. The second reason is associated to the memetic algorithm, which has the ability of search in complex and wide search spaces with good performance. This ability allows finding new combination of the model variables, which probably can produce better solutions to the problem. The combination of these features allows the agent-based memetic algorithm exploring large search spaces and integrates the different agents' decisions, improving the solutions for the distribution network, as depicted in Fig. 4.

5 Conclusions

A multi-agent model for the solution of a three-level distribution problem was proposed and implemented in this chapter. The problem is complex enough to require the use of non-traditional solution techniques. In this multi-agent system, several stakeholders of the distribution problem—such as customers, control, vehicles, hubs, and terminals, which then become agents in the model—were considered. The agents' interactions allow finding better and more realistic problem solutions.

The multi-agent model was integrated with a memetic algorithm to solve the problem. The resulting agent-based memetic algorithm was tested and compared to the solutions obtained by the Solomon Insertion Heuristic. The solutions found with the Solomon algorithm are feasible solutions that meet the capacity and time windows constraints. The proposed model solutions improve the Solomon Heuristics results, reducing not only the total distance but also the number of routes needed. This is due to the fact that the agent-based memetic algorithm allows exploring large search spaces and integrates the different agents' decisions.

The proposed agent-based memetic algorithm uses a mutation operator that facilitates exchange between terminals by customers. Through the mutation, the algorithm is able to search in wider search spaces, as every vehicle assigned to a terminal can be re-assigned to another terminal if the total distance is reduced. The search spaces are expanded with this new operator, thereby improving the algorithm solutions.

In future research, the proposal of models including several products, inventory management relationship and the interaction with other agents, as public admin-istrators, which are some of the limitations of the proposed multi agent based memetic algorithm, is likely to improve the performance of the distribution process, as well as making the model more realistic. It will be interesting to analyze the proposed multi-agent models answer efficiency in dynamic contexts such as urban freight transport with time dependent problems and dynamic routing. It will also be interesting to analyze models that integrate passengers transport (mobility) with freight transport system, coupled with the manner in which they impact some variables such as mobility and greenhouse gas emission using multi-agent micro simulation.

References

1. Bemeleit, B., Lorenz, M., Schumacher, J., Herzog, O.: Risk management in dynamic logistic systems by agent based autonomous objects. In: Dynamics in Logistics, pp. 259–266. Springer Berlin Heidelberg, Berlin, Heidelberg (2008)
2. Baindur, D., Viegas, J.M.: An agent based model concept for assessing modal share in inter-regional freight transport markets. J. Transp. Geogr. **19**, 1093–1105 (2011)
3. Baykasoglu, A., Durmusoglu, Z.D.U.: A classification scheme for agent based approaches to dynamic optimization. Artif. Intell. Rev. **41**, 261–286 (2014)
4. Sitek, P., Wikarek, J., Grzybowska, K.: A multi-agent approach to the multi-echelon capacitated vehicle routing problem. Presented at the (2014)
5. Tarimoradi, M., Zarandi, M.H.F., Zaman, H., Turksan, I.B.: Evolutionary fuzzy intelligent system for multi-objective supply chain network designs: an agent-based optimization state of the art. J. Intell. Manuf. **28**, 1551–1579 (2017)
6. Anand, N., van Duin, J.H.R., Tavasszy, L.: Framework for modelling multi-stakeholder city logistics domain using the agent based modelling approach. Transp. Res. Procedia **16**, 4–15 (2016)
7. Roorda, M.J., Cavalcante, R., McCabe, S., Kwan, H.: A conceptual framework for agent-based modelling of logistics services. Transp. Res. Part E Logist. Transp. Rev. **46**, 18–31 (2010)
8. Rousset, A., Herrmann, B., Lang, C., Philippe, L.: A survey on parallel and distributed multi-agent systems for high performance computing simulations. http://www.sciencedirect.com/science/article/pii/S1574013715300435 (2016)
9. Bozzo, R., Conca, A., Marangon, F.: Decision support system for city logistics: literature review, and guidelines for an ex-ante model. Transp. Res. Procedia **3**, 518–527 (2014)
10. Ma, Y., Zhu, J.: The study of multi-agent-based logistics system of special items. In: Proceedings of the 2nd International Conference on Computer Application and System Modeling. Atlantis Press, Paris, France (2012)
11. van Duin, J.H.R., van Kolck, A., Anand, N., Tavasszy, L. órán. A., Taniguchi, E.: Towards an agent-based modelling approach for the evaluation of dynamic usage of urban distribution centres. Procedia—Soc. Behav. Sci. **39**, 333–348 (2012)
12. Arango-Serna, M.D., Serna-Uran, C.A., Zapata-Cortes, J.A.: Multi-agent system modeling for the coordination of processes of distribution of goods using a memetic algorithm. Presented at the (2018)
13. Zapata-Cortés, J.A.: Optimización de la distribución de mercancías utilizando un modelo genético multiobjetivo de inventario colaborativo de m proveedores con n clientes. http://www.bdigital.unal.edu.co/53703/ (2016)
14. Wangapisit, O., Taniguchi, E., Teo, J.S.E., Qureshi, A.G.: Multi-agent systems modelling for evaluating joint delivery systems. Procedia—Soc. Behav. Sci. **125**, 472–483 (2014)
15. Antún, J.P., Antún, J.P., Antún, J.P., Antún, J.P.: Distribución urbana de mercancías: Estrategias con centros logísticos (2015)

16. Estrada Romeu, M.À.: Análisis de estrategias eficientes en la logística de distribución de paquetería. TDX (Tesis Dr. en Xarxa) (2008)
17. Rushton, A., Croucher, P., Baker, P.: The handbook of logistics and distribution management : understanding the supply chain
18. Crainic, T.G., Ricciardi, N., Storchi, G.: Advanced freight transportation systems for congested urban areas. Transp. Res. Part C Emerg. Technol. 12, 119–137 (2004)
19. Laporte, G.: Location-routing problems. In: North-Holland (ed.) vehicle routing: methods and studies. pp. 163–198. Amsterdam (1988)
20. Taniguchi, E., Thompson, R.G., Yamada, T., van Duin, R.: City logistics. Emerald Group Publishing Limited (2001)
21. Eltantawy, R., Paulraj, A., Giunipero, L., Naslund, D., Thute, A.A.: Towards supply chain coordination and productivity in a three echelon supply chain. Int. J. Oper. Prod. Manag. 35, 895–924 (2015)
22. Serna Urán, C.A.: Modelo multi-agente para problemas de recogida y entrega de mercancías con ventanas de tiempo usando un algoritmo memético con relajaciones difusas (2016)
23. Ehmke, J.F., Steinert, A., Mattfeld, D.C.: Advanced routing for city logistics service providers based on time-dependent travel times. J. Comput. Sci. 3, 193–205 (2012)
24. Taillard, E.D., Laporte, G., Gendreau, M.: Vehicle routeing with multiple use of vehicles. J. Oper. Res. Soc. 47, 1065 (1996)
25. Mańdziuk, J., Żychowski, A.: A memetic approach to vehicle routing problem with dynamic requests. Appl. Soft Comput. 48, 522–534 (2016)
26. Walteros, J.L., Medaglia, A.L., Riaño, G.: Hybrid algorithm for route design on bus rapid transit systems. Transp. Sci. 49, 66–84 (2015)
27. Beuck, U., Rieser, M., Strippgen, D., Balmer, M., Nagel, K.: Preliminary results of a multi-agent traffic simulation for Berlin. In: The Dynamics of Complex Urban Systems, pp. 75–94. Physica-Verlag HD, Heidelberg (2008)
28. Kickhöfer, B., Nagel, K.: Towards high-resolution first-best air pollution tolls. Networks Spat. Econ. 16, 175–198 (2016)
29. Raney, B., Cetin, N., Völlmy, A., Vrtic, M., Axhausen, K., Nagel, K.: An agent-based microsimulation model of Swiss travel: first results. Networks Spat. Econ. 3, 23–41 (2003)
30. Cevirici, A., Moller-Madsen, H.: Solving logistic problem with multi-agent system
31. Taniguchi, E., Thompson, R.G., Yamada, T.: Emerging techniques for enhancing the practical application of city logistics models. Procedia—Soc. Behav. Sci. 39, 3–18 (2012)
32. van Lon, R.R.S., Holvoet, T.: Towards systematic evaluation of multi-agent systems in large scale and dynamic logistics. Presented at the October 26 (2015)
33. Schroeder, S., Zilske, M., Liedtke, G., Nagel, K.: Towards a multi-agent logistics and commercial transport model: the transport service provider's view. Procedia—Soc. Behav. Sci. 39, 649–663 (2012)
34. Maciejewski, M., Nagel, K.: Towards multi-agent simulation of the dynamic vehicle routing problem in MATSim. Presented at the (2012)
35. Barbucha, D.: Solving instances of the capacitated vehicle routing problem using multi-agent non-distributed and distributed environment. Presented at the (2013)
36. Kumari, S., Singh, A., Mishra, N., Garza-Reyes, J.A.: A multi-agent architecture for outsourcing SMEs manufacturing supply chain. Robot. Comput. Integr. Manuf. 36, 36–44 (2015)
37. Matteis, T., Liedtke, G., Wisetjindawat, W.: A framework for incorporating market interactions in an agent based model for freight transport. Transp. Res. Procedia 12, 925–937. Elsevier (2016)
38. Arango-Serna, M.D., Serna Urán, C.A.: Un nuevo protocolo de negociación basado en inferencia difusa aplicado a la cadena de suministros. Dirección de Investigación y Postgrado, Vicerrectorado Puerto Ordaz de la Universidad Nacional Experimental Politécnica "Antonio José de Sucre" (2016)

39. Ghadimi, P., Ghassemi Toosi, F., Heavey, C.: A multi-agent systems approach for sustainable supplier selection and order allocation in a partnership supply chain. http://www.sciencedirect.com/science/article/pii/S0377221717306410 (2017)

40. De Oliveira, L.K., Lessa, D.A., Oliveira, E., Gregório Calazans, B.F.: Multi-agent modelling approach for evaluating the city logistics dynamic in a vulnerability situation: an exploratory study in Belo Horizonte (Brazil). Transp. Res. Procedia. **25**, 1046–1060. Elsevier (2017)

41. IEEE: Welcome to the Foundation for Intelligent Physical Agents. http://www.fipa.org/

42. Bellifemine, F.L., Caire, G., Greenwood, D.: Wiley InterScience (Online service). In: Developing multi-agent systems with JADE. Wiley (2007)

43. Arango Serna, M.D., Serna Uran, C.A.: A memetic algorithm for the traveling salesman problem. IEEE Lat. Am. Trans. **13**, 2674–2679 (2015)

44. Arango-Serna, M.D., Serna-Uran, C.A., Zapata-Cortes, J.A., Alvarez-Benitez, A.F.: Vehicle routing to multiple warehouses using a memetic algorithm. Procedia—Soc. Behav. Sci. **160**, 587–596 (2014)

45. Marinakis, Y., Marinaki, M.: A hybrid genetic—particle swarm optimization algorithm for the vehicle routing problem. Expert Syst. Appl. **37**, 1446–1455 (2010)

46. Tavakkoli-Moghaddam, R., Saremi, A.R., Ziaee, M.S.: A memetic algorithm for a vehicle routing problem with backhauls. Appl. Math. Comput. **181**, 1049–1060 (2006)

47. El fallahi, A., Prins, C., Wolfler Calvo, R.: A memetic algorithm and a tabu search for the multi-compartment vehicle routing problem. Comput. Oper. Res. **35**, 1725–1741 (2008)

48. Arango-Serna, M.D., Zapata-Cortes, J.A., Serna-Uran, C.A.: Collaborative multiobjective model for urban goods distribution optimization. Presented at the (2018)

49. Solomon, M.M.: Algorithms for the vehicle routing and scheduling problems with time window constraints. Oper. Res. **35**, 254–265 (1987)

Part II
Case Studies: Clinical, Emergency Management and Pollution Control

DiabSoft: A System for Diabetes Prevention, Monitoring, and Treatment

Nancy Aracely Cruz-Ramos, Giner Alor-Hernández,
José Luis Sánchez-Cervantes, Mario Andrés Paredes-Valverde
and María del Pilar Salas-Zárate

Abstract Most decision support systems for diabetes management are unable to adapt to the specific functional requirements of patients and physicians. Each physician has a different clinical experience, so she/he expects a specific decision-making scheme, and each patient has a different health plan that requires personalized care. Also, current support systems may be incapable of adapting to rapid changes in demand, especially as the change and renewal guidelines of clinical systems are often revised. To address such limitations, we introduce DiabSoft, a medical decision support system that prevents, monitors, and treats diabetes. DiabSoft relies on data interchange and integration to generate medical recommendations for patients and health care professionals.

Keywords Diabetes · Treatment · Recommender system

1 Introduction

Various diabetes management systems and programs aimed at improving glycemic control rely on information and communication technologies (ICTs), yet they offer limited or non-real-time interaction between patients and the system in terms of the system's response to patient input. Likewise, few studies have effectively evaluated the usability and feasibility of support systems to determine how well patients understand and can adopt the technology involved. The main obstacles to motivation are usability issues, self-efficacy, and the absence of clear triggers for using

N. A. Cruz-Ramos · G. Alor-Hernández (✉) · M. A. Paredes-Valverde ·
M. d. P. Salas-Zárate
Division of Research and Postgraduate Studies, Instituto Tecnológico de Orizaba,
Av. Oriente 9 no. 852 Col. E. Zapata, CP 94320 Orizaba, Veracruz, Mexico
e-mail: galor@itorizaba.edu.mx

J. L. Sánchez-Cervantes
CONACYT-Instituto Tecnológico de Orizaba, Av. Oriente 9 no. 852 Col. E. Zapata,
CP 94320 Orizaba, Veracruz, Mexico

© Springer International Publishing AG 2018
R. Valencia-García et al. (eds.), *Exploring Intelligent Decision Support Systems*,
Studies in Computational Intelligence 764,
https://doi.org/10.1007/978-3-319-74002-7_7

the applications. Therefore, to increase the adoption rates of self-management support systems, it is important to provide the right incentives, tailor technology to patient needs and requirements, and increase ease of use.

Nowadays, it is important to integrate self-management systems with existing hospital infrastructure. This can help automated systems to better analyze the wide range of data and provide adequate feedback to both healthcare professionals and patients. Also, combining self-management systems with hospital facilities would enhance the exchange of online information between nurses and physicians and between clinicians and patients. In this sense, the best way to incorporate user requirements is to engage end users in the system design and development (user-centered design) stages and to adopt rapid or agile development methods.

According to [1], diabetes mellitus (DM) is a non-communicable and non-curable everlasting disorder characterized by abnormally high blood sugar (glucose) levels. If left untreated, DM can cause a series of complications and deadly diseases, including kidney failure, stroke, heart-related diseases, cancer, and blindness. Also, because diabetes is a silent-killer disease, it demands proper patient care and sound self-management.

DM occurs for three reasons: (1) because of inadequate insulin production, (2) because the body cells are incapable of responding to insulin, or (3) because of both, (1) and (2). Similarly, there are three types of diabetes. Type 1 diabetes usually develops during childhood and before the age of 40 years old. Type 2 is the most common type of diabetes and is due to several factors. Finally, gestational diabetes is a temporary condition that develops during pregnancy. In any case, DM demands proper patient care, which includes proper medication, diet management, physical activity, knowledge of diabetes, social and individual awareness, and strict self-discipline.

The main challenge, yet the key factor, for diabetes sufferers is lifestyle management. That said, the evolution and current applications of ICTs and the Internet demonstrate that the lifestyle of diabetes patients can be guided using mobile applications. Similarly, the Web can provide multiple benefits for healthcare, such as full patient access to medical services and assistance, regardless of the patient's genre, race, or sexual orientation (i.e. people only need Internet access). Also, the Web promotes interactive opinion and information exchange through social media, forums, blogs, and open access communities, among others.

Other advantages of the Internet in the medical field include improved patient-physician relationships, channeled medical information, and patient education. More specifically, by using the Web, patients interact more with their physicians and thus attend more consults. As a result, sufferers can understand better the causes and consequences of their condition and can learn how to treat DM appropriately [2]. Likewise, with Web-based support systems, doctors can assist patients by conducting or filtering medical information, thereby enabling the patient to access much more reliable data. Finally, researching, consulting information with physicians, and receiving suggestions are forms of education. These processes ensure that patients receive the necessary quality information and increase patient autonomy in and responsibility for health management without avoiding medical suggestions and by keeping in mind that their physician can be easily reached when necessary [3].

To many patients, mobile devices with Internet access are efficient, time-saving tools. From making online appointments to seeking second medical opinions or online treatment [4]. Web-based and mobile support systems can streamline certain medical services without requiring patients to visit the doctor's office. Also, such systems allow patients to deepen into healthcare services that are truly useful to them, increase patient-doctor communication, and provide personalized services at accessible prices. In other words, ICTs applied in the medical field can guarantee that everybody have access to quality healthcare information and healthcare services anytime and anywhere. Moreover, the Web increases the organization and safety of a patient's medical history [5].

DiabSoft, as a medical decision support system, includes a Web platform for patient monitoring and a medical recommendations system. This recommendation system relies on collaborative and knowledge-based filtering, semantic technologies, data mining, and collective intelligence to prevent and treat diabetes. Also, the recommendations are generated from a knowledge base that includes information provided by the patient (vital, physical, and mental parameters) through mobile devices, medical history, medical experience, medical opinions from two social networks—Facebook and Twitter—and detected patient behavior patterns. Likewise, DiabSoft follows up patient compliance with the generated recommendations and offers feedback before and after the treatment.

2 Related Works

We conducted a comprehensive review of the literature to study current prevention and medical recommendation platforms, frameworks, systems, and applications. The research areas reviewed included eHealth, health sciences, information technologies, mobile applications, medical social networks, and emergency medicine, among others. In this sense, we found that [6] proposed a knowledge-based clinical decision support system to monitor patients suffering from chronic diseases and improve their life quality. The system allows patients to select the lifestyle changes they are willing to adopt, and such preferences are used by the system to generate personalized recommendations.

In their work, Waki et al. [7] presented DialBetics, a smartphone-based self-management support system for patients suffering from type 2 diabetes. DialBetics is composed of four modules: the data transmission module, the evaluation module, the communication module, and the dietary evaluation module. Also, the system is a real-time, partially automated interactive system that interprets patient data—biological information, exercise, and dietary content—and responds with appropriate actionable findings, thereby helping patients achieve diabetes self-management. To test DialBetics, 54 type 2 diabetes patients were randomly divided into two groups, 27 were included in the DialBetics group and 27 in the non-DialBetics group. As major findings, the authors discovered that HbA1c and fasting blood sugar (FBS) values declined significantly in the DialBetics group.

Researchers [8] introduced an electronic health record-based diabetes medical decision support system to control hemoglobin A1c (glycated hemoglobin), blood pressure, and low-density lipoprotein (LDL) cholesterol levels in adults with diabetes. The system evaluation was conducted among 11 clinics with 41 consenting primary care physicians and the physicians' 2556 patients with diabetes. The patients were randomized either to receive or not to receive an electronic health record-based system to improve care for those patients whose hemoglobin A1c, blood pressure, or LDL cholesterol levels were higher than the goal at any office visit. Among the intervention group physicians, 94% were satisfied or very satisfied with the intervention, and moderate use of the system persisted for more than one year after the authors discontinued feedback and incentives.

El-Gayar et al. [9] conducted a systematic review to determine how information technologies (ITs) have been used to improve diabetes self-management among adults with type 1 and type 2 diabetes. Overall, 74% of the studies showed some form of added benefit, 13% of the articles showed no-significant value provided by ITs, and 13% did not clearly define the added benefit due to ITs. The information technologies used included the Internet (47%), cellular phones (32%), telemedicine (12%), and decision support techniques (9%), whereas the identified limitations and research gaps ranged from real-time feedback and integration with Electronic Medical Record (EMR) provider to analytics and decision support capabilities. In the end, [9] concluded that the effectiveness of self-management systems should be assessed along with multiple dimensions, including motivation for self-management, long-term adherence, cost, adoption, satisfaction, and outcomes.

In their research, Liu et al. [10] introduced the concept, the development process, and the architecture of clinical decision support system (CDSS); then, they reviewed CDSS application progress and problems. Finally, as future directions, the authors argue that, to improve CDSS, it is important to enhance technological research, construct the knowledge base, pay attention to organizational and cultural factors, strengthen project management, improve CDSS portability, raise awareness and acceptance of healthcare workers, and improve CDSS cost-effectiveness. On the other hand, Jung et al. [11] proposed evolutionary rule decision making using similarity based associative chronic disease patients to normalize clinical conditions by utilizing information of each patient and recommend guidelines corresponding to detailed conditions in CDSS rule-based inference. The authors modified the conventional CDSS rule-based algorithm to inform of unique characteristics of chronic disease patients and preventive strategies and guidelines for complex diseases. Also, the proposed evolutionary rule decision making program selectively uses a range of database in chronic disease patients.

El Gayar et al. [12] revised a set of commercially available mobile applications to assess their impact on diabetes self-management. Overall, the review indicated that mobile applications are viable tools for diabetes self-management and are preferred over Web- or computer-based systems when it comes to usability. The review also found that diabetes self-management applications are as useful to patients as they are to providers. Also, the use of mobile applications reportedly improved positive health habits, such as healthy eating, physical activity, and blood glucose testing.

Wan et al. [13] evaluated the uptake and use of the Electronic Decision Support (EDS) tool developed by the Australian Pharmaceutical Alliance and described its impact on the primary care consultation for diabetes from the perspectives of general practitioners and practice nurses. As major findings, the qualitative analysis found that EDS had a positive impact on the quality of care of type 2 diabetes and improved the effectiveness of consultations of patients with type 2 diabetes. Similarly, Kirwan et al. [14] examined the effectiveness of a freely available smartphone application combined with text-message feedback from a certified diabetes educator. In the end, the authors found that the use of a diabetes-related smartphone application combined with weekly text-message support from a health care professional can significantly improve glycemic control in adults with type 1 diabetes. Finally, Cafazzo et al. [15] proposed to develop an mHealth application for the management of type 1 diabetes in adolescents. The application rewards the behaviors and actions of patients in the form of iTunes music and apps. As the major finding, the mHealth application with the use of incentives showed an improvement in the frequency of blood glucose monitoring in adolescents with type 1 diabetes.

Following this review of the literature, Table 1 compares the most relevant featured of the revised initiatives for diabetes prevention, monitoring, and treatment.

As can be observed, Health Information Technologies (HITs) encourage patient self-health management and relieve users from the daily frustrations of trying to adopt a normal lifestyle. The tools discussed above demonstrate important advances in the care of diabetes; however, most of these tools do not address all the phases of diabetes, have usability problems, or are not interoperable. Also, these tools struggle to integrate with other health systems and show low user acceptance due to high costs. To address such limitations, DiabSoft is conceived as a comprehensive platform that covers all the phases of chronic-degenerative diseases. Moreover, thanks to its scalable design, the usability of DiabSoft can be extended by including more diseases into the architecture's integration layer.

The goal of DiabSoft is to be a personalized recommendation tool that adapts to each patient's needs. To reach this goal, the system combines different ITs, such as ontologies, big data, data mining, recommendation systems, opinion mining, and social media, among others. Also, DiabSoft offers a Web platform that is accessible through a mobile application to monitor the disease status through a set of parameters (vital, physical and mental), physical activities, and the patient's medical history.

DiabSoft's disease monitoring approach includes alerts and reminders regarding when the parameters should be provided and how often (based on the knowledge defined by the developed ontologies). Once the recommendations are generated and the patient accepts them, DiabSoft provides a follow-up program for the patient to follow the health recommendations. Considering the self-management philosophy, DiabSoft users can follow up on their treatment through their mobile devices, obtain

Table 1 Comparison of systems initiatives for the prevention, monitoring and treatment of diabetes

Author	Technologies	Diseases	Stage
Vives-Boix et al. (2017) [6]	Not specified	Diabetes Hypertension	Treatment Follow-up
Waki et al. (2014) [7]	NLP method Wi-Fi Bluetooth NFC	Diabetes	Monitoring Treatment
O'Connor et al. (2012) [8]	Clinical algorithms Linear mixed models	Diabetes Hypertension Cardiovascular diseases Coronary heart diseases	Monitoring Treatment Follow-up
El-Gayar et al. (2013) [9]	Decision support techniques Telemedicine Data mining	Diabetes	Follow-up Treatment
Liu et al. (2016) [10]	Decision support techniques	Diabetes Hypertension	Diagnosis Monitoring Treatment
Jung et al. (2015) [11]	U-Healthcare services Data mining Decision support techniques Telemedicine IT convergence NLP models SOAP EMR	Hypertension Hyperlipidemia Diabetes	Diagnosis Prevention Monitoring
El Gayar et al. (2013) [12]	Decision support techniques iOS Social media IoT Data mining	Diabetes	Monitoring Treatment
Wan et al. (2012) [13]	Not specified	Diabetes	Follow-up Treatment
Kirwan et al. (2013) [14]	Mobile technologies Linear mixed effects models	Diabetes	Monitoring Treatment Follow-up
Cafazzo et al. (2012) [15]	iOS Bluetooth Social media	Diabetes Hypertension	Monitoring Treatment

timely information and notifications, change the treatment plan, and record changes in their health status or any other information that is valuable to keep track of their recovery. Also, DiabSoft users are informed if any of the changes they perform causes a setback in the treatment. Moreover, through its collaborative environment, DiabSoft offers users feedback and recommendations previously filtered and valued by other patients and health professionals. Likewise, because social networks currently play a key role in the mental state of patients with chronic diseases, one of the platform's modules integrates information from specialized social networks for certain diseases, where patients share their experiences. Finally, DiabSoft's modules are based on ontologies and archetypes that model knowledge and clinical examinations to adapt to different diseases.

3 Architecture of DiabSoft System

The rise of digital technology and the Internet have paved the way for innovations that make life easier and more sustainable. One of these technologies is eHealth, a broad term that groups health practices assisted by digital and mobile technology. EHealth has great benefits for users either in the role of patients, healthcare specialists, health organizations, or governments. DiabSoft seeks to contribute to the benefits of eHealth through an innovative architecture that includes the following characteristics:

(a) **Web platform**: DiabSoft's platform includes three elements. First, it includes the chronic-degenerative diseases classification Ontology (RDF and OWL) and the medical services classification Ontology (RDF and OWL) to adapt the system and specialize the modules inexpensively to treat different diseases. Second, DiabSoft has an eHealth platform with a monitoring module to supervise not only the patient's vital, physical, and mental parameters, but also the physical and mental activities performed. The module also ensures the parameters' transparent, coherent, and standard-based integration in the patient's electronic medical records. This platform will be available as a responsive Web application, a Web application optimized for mobile devices, and as an application for Android-based mobile devices. Finally, DiabSoft relies on a knowledge-based and collaborative-filtering-based recommendation system.

(b) **System of recommendations, treatment, and patient follow-up**: This system includes three elements: a medical opinions repository, a medical opinions extractor, and a treatment tracking system. The medical opinions repository helps treat chronic-degenerative diseases using medical-related information extracted from social networks where healthcare specialists actively participate. Also, this repository relies on Facebook and Twitter, the two most popular social networks. On the other hand, the medical opinions extractor obtains information from the abovementioned social networks and stores it in the medical opinions repository. Finally, the treatment tracking system provides traceability of the generated medical recommendations to identify and inform users of the health impact of such recommendations. This system continuously improves the recommendations generated through the informative feedback that is provided to healthcare specialists. Moreover, DiabSoft's multi-device interface supports iOS and Android mobile operating systems.

(c) **Data Exploitation and Clinical Validation** through an opinion mining system, which exploits user-generated information available in the platform's repositories. Also, DiabSoft diagnoses diseases, recommends treatments, and provides follow-up not only based on similarity or experiences of previous clinical history, but also according to user behavior patterns and depending on the type of disease, its stage (e.g. terminal stage), the patient's mood, and other social factors retrieved from medical social networks. Likewise, DiabSoft has a computer service that is sensitive to the medical context. DiabSoft integrates with other medical systems, operates with new platforms, and supports different types of devices through an integration layer based on interoperability and semantic mapping technologies.

As can be observed, DiabSoft is a scalable system for clinical knowledge management that allows users to access medical information in a timely manner. The system thus improves decision making, correlations, indicators definition, and timely detection of chronic-degenerative diseases. Moreover, DiabSoft generates medical recommendations and integrates with social media to support opinion mining techniques, collective intelligence, and the use of semantic technologies such as Web Ontology Language (OWL) and RDF (Linked Data/Open Data) data. As a result, the system can generate medical recommendations that support patients in their efforts to prevent, monitor, and treat diabetes. Finally, DiabSoft relies on a layered architecture to offer scalability, robustness, and easy maintenance. This layered architecture and its modules are depicted in Fig. 1.

Presentation Layer: This layer receives user requests entered through different means of access. Namely, the presentation layer offers support for a multi-device application, which provides user access through a browser (cross-browser, Chrome, Safari, and Firefox) for the most popular operating systems (e.g. Windows, macOS X, and Linux). Moreover, DiabSoft's native version (mobile client) is available as a mobile application for Android, iOS, and Windows Phone. However, given the modularity and scalability of the architecture, the capabilities of the presentation layer can be extended to other platforms and devices, since the lowest integration layer provides a service-based application programming interface (API).

Integration Layer: This layer offers an API to access diverse functionalities of the lowest layers, and as a result, it offers low coupling between the user and the system's functionality. Broadly speaking, this layer allows DiabSoft to create new clients by adding a new platform or device support through the presentation layer. Also, the integration layer serves as both a link and isolation and a security mechanism to access the processes and functions of the service layer. Likewise, this layer includes the access to the four main diabetes management services: Prevention, Treatment, Monitoring, and Follow-up.

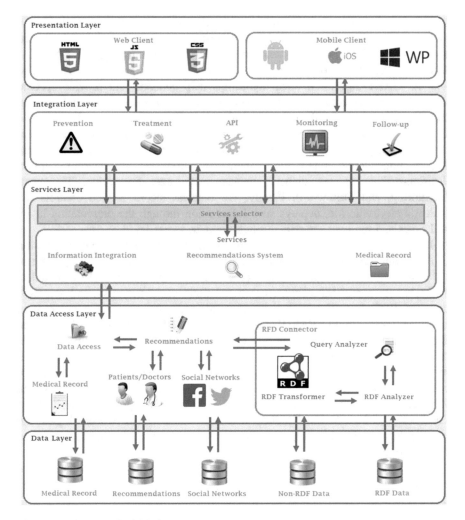

Fig. 1 Architecture of DiabSoft

Services Layer: This layer has a module that selects the requested services and validates the parameters sent by the integration layer. Also, the services layer ensures the request origin and the credentials to access the data from lower layers. There is also a set of services whose main function is to control the system's services. Likewise, this layer includes the services selector component to select the requested services and validate the parameters sent by the integration layer to either grant or deny a service according to the received parameters and the used authentication credentials.

Additionally, the services layer includes three services: information integration, recommendation system, and medical record. The information integration service is an orchestrator of information from different data sources, and it anonymizes the sensitive data used in recommendations. On the other hand, the recommendation system uses different recommendation algorithms, including decision trees, the K-means algorithm, support vector machines, the Apriori algorithm, the EM algorithm, PageRank, AdaBoost, the k-nearest neighbors algorithm, Naive Bayes, and the CART algorithm. Such algorithms help generate and provide specific medical recommendations to treat chronic-degenerative diseases. Also, data mining techniques are applied to the medical opinions repository to analyze and summarize the data from different perspectives, identify patterns and relationships, and transform data into proactive knowledge exploited through the recommendation system. The recommendation system gives patients and physicians information to support decision-making at any diabetes management stage: Prevention, Treatment, Monitoring, and Follow-up). Finally, the medical record service controls the patient's electronic file, deploys such information through the presentation layer, and saves and retrieves the data through the data access layer.

Data Access Layer: This layer grants access to all the input/output information of the architecture and facilitates the access to the various data sources present in the data layer. To this end, the data access layer invocates the five access modules: RDF connector, medical record, recommendations, social networks, and non-RDF data and RDF data. The RDF connector uses the query analyzer component to transform requests into SPARQL queries to access RDF and non-RDF data from the RDF transformer component. On the other hand, the medical record module refers to the patient's electronic medical history and monitors the patient's vital and physical parameters and physical and mental activities. The recommendations are based on the user (patient), historical medical diagnoses, and social networks. In turn, the social networks access module is a medical opinions repository of Facebook and Twitter. Finally, RDF data are obtained from two ontologies: the chronic-degenerative diseases classification ontology and the medical services classification ontology.

Data Layer: This layer stores all the data—structured data, unstructured data, and social networks data—used by the upper layers. Simultaneously, the data layer can add new data sources into the system.

3.1 Overview of the Architecture Based on Its Functionality

The following paragraphs discuss the modules of DiabSoft and explore the functionality of the architecture:

Repository of knowledge bases, ontologies, and linked data: This module manages and provides a common interface to store information and knowledge in OWL and RDF formats, which allow the system to access and publish knowledge bases through linked data. Each knowledge base has a T-Box layer and an A-Box layer. The T-Box layer describes the structure of the classes and the relationships and constraints of the knowledge base, whereas the A-Box layer includes the knowledge base instances. This repository provides the system with reasoning capabilities to infer new knowledge from pre-existing information in the different knowledge bases.

Module of transparent and intelligent clinical information integration: The final goal of this module is to construct a system of data (including clinical information standards) portability, interchange, interoperability, and integration. To this end, the architecture uses interoperability and semantic technologies as ontologies. This module is an orchestrator of the information that comes from the multiple data sources. Also, it anonymizes sensitive data used in recommendations and controls the patient's electronic file to detect whether the information to be incorporated into the patient's medical record is insufficient or incomplete.

Module of monitoring and alert management: This module monitors the condition of each patient through input parameters. Such parameters are entered by the patient himself/herself using a mobile device. The monitoring approach includes alerts and reminders regarding when and how often the parameters must be entered, according to the knowledge defined by the developed ontologies.

Intelligent recommendation module: Using all the information introduced by the patient and medical professionals, this module generates relevant recommendations based on expert knowledge, such as disease guidelines. These guidelines are obtained from information provided by other patients and health professionals. Each patient has a personal profile that stores user parameters, health condition, and current treatment. Finally, by using knowledge-based technologies and data mining, the system also tries to locate interesting health recommendations and content.

Treatment follow-up and recommendations module: Once the recommendations are generated and accepted, the system launches the follow-up care for the patient to follow such health recommendations. This module offers various follow-up alternatives; patients can keep track of their treatment through their mobile devices, obtain timely information and notifications, edit the treatment plan, and change health records or any other information that is valuable to keep track of their recovery. Also, DiabSoft informs users whether a given change in their lifestyle causes a retreat in the treatment. Additionally, thanks to DiabSoft's collaborative environment, patients can obtain feedback and recommendations previously filtered and valued by other patients and healthcare professionals. Moreover, because social networks play a key role in the mental health of patients, this module integrates with disease-specialized social networks for patients to share their experiences.

Multi-device user interface: This interface allows the user (patient or healthcare professional) to control the system's functionality and identifies different types of users, such as module administrators, consumers, and users.

Finally, note that the system's modules rely on ontologies and archetypes that model knowledge and clinical exams to adapt inexpensively to different diseases. For instance, the medical services classification ontology is based on the traditional ontological model of concepts, relationships, attributes, and axioms. Also, DiabSoft has a special module for diabetes, although eventually, the system can also include special modules for arterial hypertension and hepatitis.

4   Case Study

For this case study, a patient has to monitor their health to prevent DM occurrence. More specifically, the patient's needs are:

(a) Monitor their habits, symptoms, and vital signs to keep track of their health and thus prevent DM occurrence.
(b) Obtain recommendations to prevent DM.

As an alternative tool to meet their needs, the patient uses DiabSoft to record and consult their symptoms, habits and vital signs.

4.1   User Monitoring

DM monitoring is one of the key parts of the referral process of DiabSoft. In this sense, the user enters the necessary information for the system to keep track of their health. The information stored in the system is not only useful to the user (i.e. patient), but also to their physician(s).

4.1.1   Record of Symptoms, Vital Signs and User Habits

As shown in Fig. 2, after logging into DiabSoft, the registered users can select one of the six options: My Profile, My Monitoring, My Consults, My Habits, Recommendations, or Doctors. Each section provides specialized information adapted to the needs of the patient.

The patient selects My Monitoring. To obtain the necessary data to monitor their health parameters, the patient clicks on Diabetes, and DiabSoft loads the symptoms

Fig. 2 Main menu of a registered DiabSoft user

Fig. 3 Symptoms record

Fig. 4 Vital signs record

registered by the user in Symptoms Record. This process is depicted in Fig. 3. Note that the user can eliminate some of the registered symptoms when desired. To this end, DiabSoft provides a keyword search option to simplify user-system interaction.

Figure 4 shows how the patient can record their vital signs in the Vital Signs Record section. The signs registered by the user are then displayed. In this section, DiabSoft also provides a keyword search mechanism to facilitate the patient's interaction with the system.

To register the symptoms, the patient must select at least one of the suffered symptoms. Then, they must click on the "Save changes" button to store the selected item or items (i.e. symptoms). As depicted in Fig. 5, these symptoms can be updated at any time, depending on how the patient's health status evolves.

To record their vital signs, the user must enter the data in the displayed form and click on the "Save changes" button as seen in Fig. 6. The vital signs are registered on awakening (fasting parameters) and two hours after a meal (postprandial parameters).

Figure 7 shows the medical record of habits entered and saved by the patient. The registered habits can also be removed as the patient's lifestyle changes. As in previous cases, DiabSoft offers a search tool to expedite the habits search.

As Figs. 3, 4, and 7 depict, the patient can keep track of their disease and is responsible for recording the necessary data. Such information is extremely important to DiabSoft to generate the appropriate medical recommendations. Also, to register their habits, the patient must select their common habits. As shown in

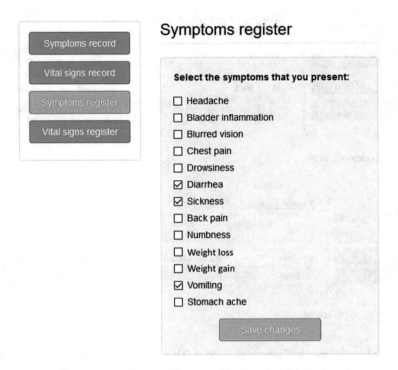

Fig. 5 Symptoms register

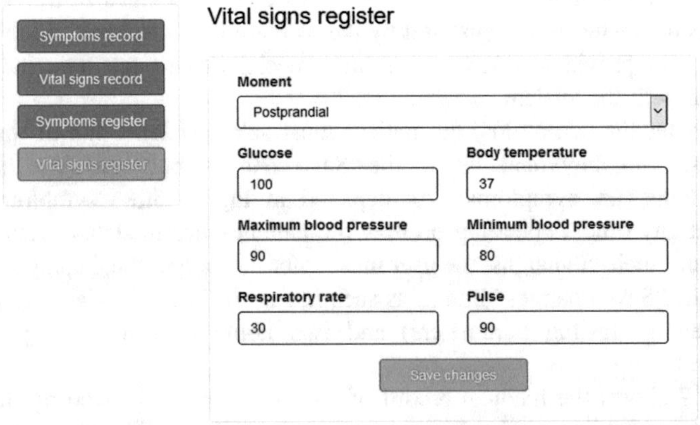

Fig. 6 Vital signs register

Fig. 8, at the end of the process, the user must click on the "Save changes" button to store the selected items. These habits can be updated at any time, but they can be registered only once a day.

Fig. 7 Habits record

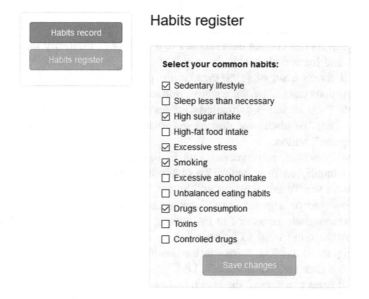

Fig. 8 Common habits register

4.1.2 Recommendations for Diabetes Prevention and Care

DiabSoft generates medical recommendations through collaborative filtering and helps users manage various health aspects. The first recommendations that patients must consider are those related to their daily habits. By changing their current

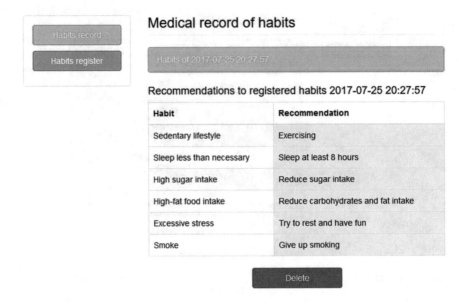

Fig. 9 Recommendations of system to user habits

lifestyle, patients can control the maladies that directly or indirectly impact on the vital signs and harm their health.

Figure 9 shows a set of recommendations provided by DiabSoft. If followed, these suggestions can reduce the negative impact of the patient's current lifestyle on their health. Such impact is particularly reflected through the symptoms registered on the platform. To obtain the desired information, the patient must click on the "Habits register" button.

DiabSoft generates recommendations to improve the vital signs of diabetic patients by modifying their habits. To obtain these recommendations, the patient must click on the "Vital signs record" button. The recommendations are concise and are displayed on a green or red background. The suggestions presented on a green background are beneficial to the patient, whereas the suggestions displayed on a red background need to be attended by a specialist or the patient himself/herself. Figure 10 depicts some of the health recommendations generated by DiabSoft to a user seeking to prevent DM.

DiabSoft users can consult the health records of their symptoms, vital signs, and habits. To obtain this information, patients first must click on the "Register habits" button. Then, DiabSoft provides an available link, such as "Check your health records." Figures 11 and 12 depict this functionality.

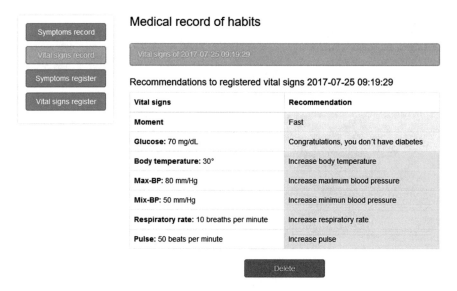

Fig. 10 Vital signs recommendations to prevent diabetes

Fig. 11 Recommendations for users with diabetes

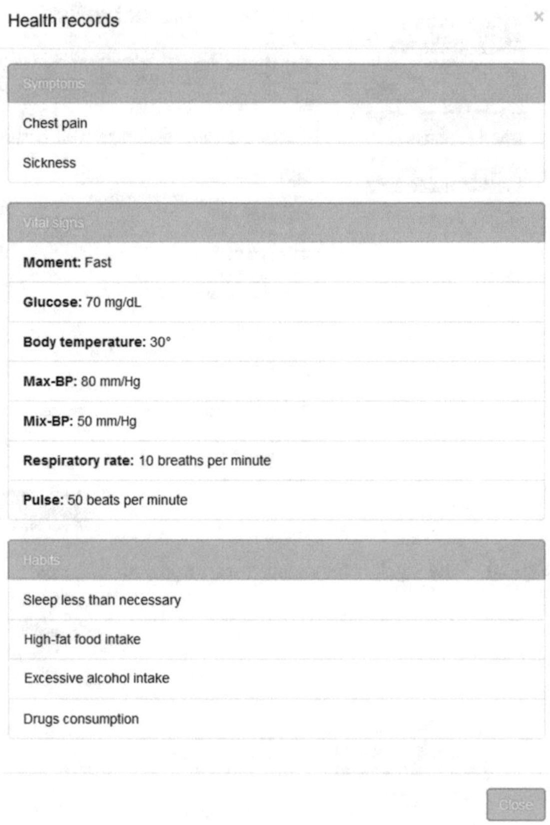

Fig. 12 Deployment of
information on symptoms,
vital signs, and habits of users
with diabetes

Health records   ×

Symptoms

Chest pain

Sickness

Vital signs

Moment: Fast

Glucose: 70 mg/dL

Body temperature: 30°

Max-BP: 80 mm/Hg

Mix-BP: 50 mm/Hg

Respiratory rate: 10 breaths per minute

Pulse: 50 beats per minute

Habits

Sleep less than necessary

High-fat food intake

Excessive alcohol intake

Drugs consumption

Close

5   Conclusions and Future Directions

Current systems for diabetes prevention, monitoring, and treatment along with the
use of state-of-the-art technologies significantly improve the quality of life of
people with diabetes. This work proposes DiabSoft, an innovative system for
preventing, treating, and monitoring diabates. DiabSoft provides healthcare specialists and public and private health institutions with a tool to support medical
decisions to prevent, diagnose, treat, and follow-up chronic-degenerative diseases
such as diabetes. Also, DiabSoft monitors patient health parameters mainly to
obtain data on their vital signs, daily habits, and symptoms, and then generate
recommendations based on collaborative filtering and shared knowledge. Such
strategies help prevent and treat DM.

The case study presented in this chapter demonstrates the system's simplicity.
The information stored in DiabSoft is intended to be useful not only to patients with
diabetes, but also to the physicians and healthcare specialists who care for them.
Similarly, user (i.e. patient) data stored in DiabSoft can help prevent other diseases,

such as hypertension. Therefore, as future work, we look forward to expanding the system's prevention capabilities to cater for other diseases. In addition, we will seek to optimize DiabSoft, so the data acquisition is carried out automatically through an interface that would link wearables with DiabSoft.

Also, we will seek to develop a medical opinions repository based on health-related and medical-related information extracted not only from social networks where healthcare specialists actively participate, but also from the most popular social networks Twitter and Facebook. Similarly, to expand the benefits of the opinions repository, a follow-up system will be developed to trace the medical recommendations generated, identify the impact of such recommendations on patient health, and inform users of such impact. Likewise, this follow-up system would improve the generated recommendations thanks to the informative feedback that DiabSoft would provide to healthcare specialists. Then, the system could identify user behavior patterns to recommend treatments, diagnoses, or follow-up care, not only based on similarity or experiences of previous clinical history, but also according to user behavior patterns, depending on the type of disease, patient mood, and other social factors identified from medical social networks. Finally, we will seek to integrate DiabSoft with other medical systems.

Acknowledgements The authors are grateful to the National Technological Institute of Mexico (Tecnológico Nacional de México) for supporting this work. This research paper was also sponsored by Mexico's National Council of Science and Technology (CONACYT) and the Secretariat of Public Education (SEP) through the PRODEP program.

References

1. Anowar, F., Shahriar Khan, M., Nurul Huda, M., Mamun, K., Bashar, M.D., Nomami, H., Nowrin, G.: A Review on diabetes patient lifestyle management using mobile application (2015)
2. Smailhodzic, E., Hooijsma, W., Boonstra, A., Langley, D.J.: Social media use in healthcare: a systematic review of effects on patients and on their relationship with healthcare professionals. BMC Health Serv. Res. **16**, 442 (2016)
3. Yang, Z., Zhou, Q., Lei, L., Zheng, K., Xiang, W.: An IoT-cloud based wearable ECG monitoring system for smart healthcare. J. Med. Syst. **40**, 286 (2016)
4. Demski, H., Garde, S., Hildebrand, C.: Open data models for smart health interconnected applications: the example of openEHR. BMC Med. Inform. Decis. Mak. **16**, 137 (2016)
5. Zhang, Y., Chen, M., Leung, V.C.M., Lai, R.: Xiaorong": topical collection on "Smart and Interactive Healthcare Systems. J. Med. Syst. **41**, 121 (2017)
6. Vives-Boix, V., Ruiz-Fernández, D., de Ramón-Fernández, A., Marcos-Jorquera, D., Gilart-Iglesias, V.: A knowledge-based clinical decision support system for monitoring chronic patients. Presented at the June (2017)
7. Waki, K., Fujita, H., Uchimura, Y., Omae, K., Aramaki, E., Kato, S., Lee, H., Kobayashi, H., Kadowaki, T., Ohe, K.: DialBetics. J. Diabetes Sci. Technol. **8**, 209–215 (2014)
8. O'Connor, P.J., Sperl-Hillen, J.M., Rush, W.A., Johnson, P.E., Amundson, G.H., Asche, S.E., Ekstrom, H.L., Gilmer, T.P.: Impact of electronic health record clinical decision support on diabetes care: a randomized trial. Ann. Fam. Med. **9**, 12–21 (2011)

9. El-Gayar, O., Timsina, P., Nawar, N., Eid, W.: A systematic review of IT for diabetes self-management: are we there yet? Int. J. Med. Inform. **82**, 637–652 (2013)
10. Liu, H., Hu, H., Chen, Q., Yu, F., Liu, Y.: Application of the clinical decision support systems in the management of chronic diseases. In: 2016 3rd International Conference on Systems and Informatics (ICSAI), pp. 482–486. IEEE (2016)
11. Jung, H., Yang, J., Woo, J.-I., Lee, B.-M., Ouyang, J., Chung, K., Lee, Y.: Evolutionary rule decision using similarity based associative chronic disease patients. Cluster Comput. **18**, 279–291 (2015)
12. El-Gayar, O., Timsina, P., Nawar, N., Eid, W.: Mobile applications for diabetes self-management: status and potential. J. Diabetes Sci. Technol. **7**, 247–262 (2013)
13. Wan, Q., Makeham, M., Zwar, N.A., Petche, S.: Qualitative evaluation of a diabetes electronic decision support tool: views of users. BMC Med. Inform. Decis. Mak. **12**, 61 (2012)
14. Kirwan, M., Vandelanotte, C., Fenning, A., Duncan, M.J.: Diabetes self-management smartphone application for adults with type 1 diabetes: randomized controlled trial. J. Med. Internet Res. **15**, e235 (2013)
15. Cafazzo, J.A., Casselman, M., Hamming, N., Katzman, D.K., Palmert, M.R.: Design of an mHealth App for the self-management of adolescent type 1 diabetes: a pilot study. J. Med. Internet Res. **14**, e70 (2012)

Health Monitor: An Intelligent Platform for the Monitorization of Patients of Chronic Diseases

José Medina-Moreira, Oscar Apolinario,
Mario Andrés Paredes-Valverde, Katty Lagos-Ortiz,
Harry Luna-Aveiga and Rafael Valencia-García

Abstract A chronic disease is a condition of long duration and generally slow progression. This kind of disease is permanent and leaves residual disability. Chronic diseases require special training of the patient for rehabilitation and require a long period of supervision, observation, and care by part of general practitioners and other healthcare professionals. This task represents a great investment that may come unbearable to healthcare organizations, especially when the number of patients increases day by day. Therefore, it is necessary to promote a healthy lifestyle among patients to improve their quality of life as well as to reduce the investment needed to treat these patients. Hence, the main goal of this work is to present Health Monitor, a platform for the self-management of chronic diseases, more specifically for diabetes mellitus and hypertension. Health monitor allows patients to monitor aspects such as physical activity, diet, medication, and mood. Also, this platform provides health recommendations based on the data collected by the motorization process. Health Monitor was evaluated regarding the effectiveness of the health recommender module. This evaluation involved real patients with diabetes mellitus and hypertension and healthcare professionals from an Ecuadorian

J. Medina-Moreira (✉) · O. Apolinario · K. Lagos-Ortiz · H. Luna-Aveiga
Cdla. Universitaria Salvador Allende, Universidad de Guayaquil, Guayaquil, Ecuador
e-mail: jose.medinamo@ug.edu.ec

O. Apolinario
e-mail: oscar.apolinarioa@ug.edu.ec

K. Lagos-Ortiz
e-mail: katty.lagos@ug.edu.ec

H. Luna-Aveiga
e-mail: harry.lunaa@ug.edu.ec

M. A. Paredes-Valverde
Division of Research and Postgraduate Studies, Instituto Tecnológico de Orizaba,
Av. Oriente 9 no. 852 Col. E. Zapata, CP 94320 Orizaba, Veracruz, Mexico
e-mail: mapv1015@hotmail.com

R. Valencia-García
Facultad de Informática, Universidad de Murcia, Campus Espinardo, 30100 Murcia, Spain
e-mail: valencia@um.es

© Springer International Publishing AG 2018
R. Valencia-García et al. (eds.), *Exploring Intelligent Decision Support Systems*,
Studies in Computational Intelligence 764,
https://doi.org/10.1007/978-3-319-74002-7_8

hospital. The evaluation results show that Health Monitor provides recommendations that can enhance patients' quality of life.

Keywords Chronic disease · Health self-management · Recommender system

1 Introduction

A chronic disease is a condition of long duration and generally slow progression [1]. According to the WHO (World Health Organization), chronic diseases are permanent, leave residual disability or are caused by nonreversible pathological alteration. Some examples of these diseases are diabetes mellitus and hypertension. Diabetes mellitus occurs when the amount of glucose in the blood is too high. This health condition occurs either when the body cannot effectively use the insulin it produces or when the pancreas does not produce enough insulin [2]. High levels of glucose in the blood may cause life-threating and disabling health complications [3]. Furthermore, the Global Report on Diabetes determined, in April 2016, that the number of patients with diabetes mellites rounds to 422 million in 2014 [2]. On the other hand, hypertension, also known as high blood pressure, occurs when the blood pressure increases to unhealthy levels, which can damage blood vessels, increase the risk for stroke and coronary heart disease, and cause other health problems. The WHO estimates that hypertension causes 7.5 million deaths, about 12.8% of the total of deaths.

As can be observed from the previous paragraph, chronic diseases are the leading causes of death and disability worldwide [4]. People with a chronic disease are managed by general practitioners and other health care professionals. This fact represents a great investment, both in terms of required money and time, that may become unbearable to healthcare organizations, especially when the number of patients increases day by day. In this sense, healthcare organizations have a need for improving financial management, operational efficiency as well as for engaging consumers in managing their health and care [5].

Chronic diseases require special training of the patient for rehabilitation and require a long period of supervision, observation or care [6]. Even though a good treatment cannot cure this kind of diseases, it can modify the disease course. Therefore, it is necessary to promote a healthy lifestyle through a better diet and the increasing of physical activity, to improve the quality of life of people with chronic disease as well to reduce the investment needed to treat these patients. Furthermore, it is necessary to reduce the gap existing between patients and health care professionals i.e., it is important to improve the communication among patients and doctors in such way that healthcare professionals have updated information about the health status of their patients and then they can establish new methods for treating the chronic disease.

Recommender systems have been successfully applied in contexts such as Web services [7], digital libraries [8], passengers and vacant taxis [9], movie show times

[10], among others. This kind of systems represents a great area of opportunity for supporting different tasks involved in the chronic disease self-management. For instance, they can provide patients with healthcare recommendations about disease treatments, food, physical activities, to mention but a few, that have proved to be effective in patients with the same disease.

On the other hand, thanks to the fast spreading of mobile devices such as smartphones and tablets the development of mobile applications, also known as apps, has increased in the last years. Nowadays, Apple's App Store and Google Play, which are the two most important mobile applications stores, offer more than 800,000 apps [11]. The use of mobile devices has transformed many aspects of different clinical practices [12] such as health record maintenance [13, 14], clinical decision-making [15, 16] and medical training and education [17, 18]. Therefore, the use of this type of applications must be considered for supporting activities related to the chronic diseases treatment such as health parameters monitorization and control as well as the generation of health care recommendations.

Considering the above-discussed, the main aim of this work is to present the design and architecture of Health Monitor, a platform for chronic diseases self-management, more specifically for diabetes mellitus and hypertension. Health Monitor allows patients to monitor different aspects of their lifestyle and disease treatment such as physical activity, diet, medication and mood. This task is performed through a mobile application. Health Monitor provides a set of recommendations based on the monitorization data collected through the mobile application as well as the similitude among patients. It is important to mention that all recommendations provided by the system were established by a group of healthcare professionals. Also, the mobile application provided by the platform allows healthcare professionals to know the current health status of their patients to help them to determine the main factors that cause it.

The rest of this work is structured as follows. Section 2 presents a review of the literature about systems for healthcare management. Section 3 presents the architecture design of Health Monitor, a platform for self-management of chronic diseases proposed in this work. Section 4 presents the evaluation performed to measure the effectiveness of the platform to provide recommendations to people with chronic diseases, more specifically, diabetes. Finally, Sect. 5 presents our conclusions and depicts future directions.

2 State of the Art

Mobile applications are very useful for several aspects of people's lives. Even, chronic diseases, such as diabetes mellitus and hypertension, can be manageable with mobile apps. Nowadays, there are several research efforts that aim to provide applications that help people with a chronic disease to control and manage different aspects related to their health and their disease's treatment including physical and psychological health. This section provides an overview of a set of health care systems that provide mobile applications for supporting patients with chronic diseases such as diabetes mellitus, pulmonary and cardiovascular diseases.

In the context of the control and management of diabetes mellitus, there are research efforts aimed at supporting these activities. For instance, in [19], the authors presented a Cloud-based system for the lifestyle management of people with Type-2 diabetes mellitus. This system is composed of a sensor network that collects data from the human body, physical activity of the patient and environmental information that may produce changes in the health status of the patient. Based on the information collected and a rule algorithm, the system provides useful information to the patients. On the other hand, E-Health Intervention [20] is a mobile application whose main objective is to serve as a coach for patients with Type-2 diabetes mellitus. E-Health Intervention allows patients to register data related to blood glucose levels, medication, activity levels and food intake. Furthermore, this mobile application analyzes input data to provide messages to patients aiming to teach and train them regarding the management of their disease.

In [21], a theoretical framework for diabetes mellitus self-management was proposed. This framework is based on a mobile application that allows patients managing three axes of diabetes treatment namely self-management (blood glucose, dietary and medication management), social support (glucose advices, food annotation) and health related outcomes (hbA1c, adherence to treatment regimen, health management success). Another application for diabetes self-management is MyDDiary [22], which covers basic aspects of diabetes treatment such as glycemic index, insulin ingestion and calorie count. The most outstanding functionality of MyDDiary is the TI (Territorial Intelligence) which allows establishing relationships among patients, territorial communities and services. These relationships are based on the events and activities planned within a personal diary that can be managed through the application. Still on the subject of systems for diabetes self-management, in [23] authors presented a RIA (Rich Internet Application) oriented to people with diabetes mellitus. This application allows tracking blood glucose test results, importing data from Microsoft HealthVault and Google Health and visualizing blood sugar levels over a period as well as compare different periods.

There are healthcare self-management systems that are designed to provide support to a specific group of patients with a chronic disease. For instance, in [24], authors proposed a set of design practices and functionalities that must have mobile technologies to support the self-management of older adults with type 2 diabetes mellitus. The design practices were obtained from an analysis of existing mobile applications available on the Android market [25] such as Glucool Diabetes, OnTrack Diabetes and Dbees.com. On the other hand, in [26], the authors presented the design and development of a mobile application focused on patients with newly diagnosed type 2 diabetes mellitus. This app allows patients to record health data and psychological data such as fatigue, well-being, and sleep. The app was evaluated by patients that found that it is a valuable tool for supporting their psychological health and for guiding them after diagnosis.

There are research works focused on the self-management of chronic diseases other than diabetes. For example, mHealth [27] is a mobile telehealth application that provides support to patients with COPD (Chronic Obstructive Pulmonary Disease). This application monitors symptoms and physiological variables as well as generates alerts when pulse rate, oxygen saturation levels and symptoms scores related to the patient's treatment are exacerbated. On the other hand, MyPADMGT [28] is an application for PAD (Peripherical Arterial Disease) self-management that is able to generate an structured report that represents the patient progress towards objectives previously stablished. Furthermore, MyPADMGT aims to educate patients about how to deal with problems related to their disease. Regarding chronic heart diseases, EasyCare [29] provides an SMS (Short Message Service), Web portal and an IVR (Interactive Voice Response) through which patients can report their health conditions to heart failure clinic. Then, care specialists visualize the patient's condition and provides advices by using SMS or email.

There are some research efforts that aim to provide applications for the continuous self-management of different chronic diseases. For instance, in [30] the authors presented the design and evaluation of SmPHR (Self-Management mobile Personal Health Record), an Android application that aims to support interoperability of personal health devices (PHDs) and electronic medical records (EMR). According to an evaluation performed, SmPHR allows plug and play of PHDs and EMRs thanks to the use of several standard protocols. On the other hand, in [31], the authors propose a mobile service based on PHR (Personal Health Record) for monitoring of high pressure, hyperlipidemia and diabetes mellitus diseases. Furthermore, this service uses IoT (Internet of Things) solutions such as smart bands and smartwatches to obtain information about physical activity of patients. myFitnessCompanion [32] is an Android application for chronic disease management that allows monitoring physiological data such as blood pressure, blood glucose, weight and heart rate, as well as oxygen, respiration, HbA1c, temperature and cholesterol. Also, this application uses Microsoft HealthVault, a PHR system.

To the best of our knowledge, there are few works that provide support for the control and self-managing of more than one chronic disease. The platform presented in this work provides support for the control and management of three different chronic diseases namely diabetes mellitus and hypertension. Furthermore, unlike most systems for the self-management of diseases analyzed in this section, Health Monitor contains a recommender module that provides patients with a set of health care recommendations about different elements of their treatment such as food and physical activities. For this purpose, this module implements a collaborative filtering method that captures the similarities among patients registered in the platform. Another outstanding feature of Health Monitor is that it allows patients to track their mood. This feature is very important because many patients that must take medication every day for the rest of their lives can suffer continuous mood swings or feel depressed. Finally, Health Monitor allows health professionals to visualize a summary of their patients' health parameters, medication, food intakes and mood. This functionality aims to help healthcare professionals to determine all those factors that have influenced the success or failure of the treatment and then generate a plan of action focused on the improvement of the patient's health's status.

3 Health Monitor's Architecture

The main objective of this work is to provide an intelligent and collaborative platform for health self-management, more specifically, for the control and management of chronic diseases such as diabetes mellitus and hypertension. The functional requirements of this platform, called Health Monitor, were acquired from a group of healthcare professionals and patients of the three chronic diseases above-mentioned. Furthermore, a group of patients with such diseases was involved in the design process of the platform. These patients, which were not involved in the evaluation process described in Sect. 4, provided a constant feedback along the development of the mobile application. Regarding the methodology used for the development of this platform, the agile methodology SCRUM [33] was selected since it provides support for developing software systems where business conditions change constantly such as is the case of this health self-management platform. The functional architecture of Health Monitor is shown in Fig. 1.

The architecture of Health Monitor (see Fig. 1) is based on a client-server application model in which the different tasks are distributed among the services' providers and the clients. The communication among server and client is performed

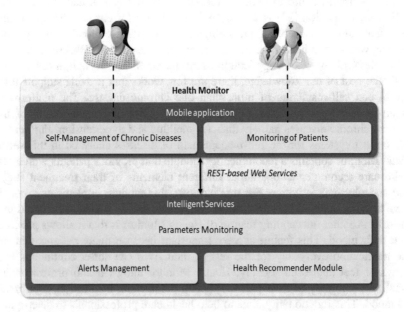

Fig. 1 General architecture of the platform for the self-monitoring of chronic diseases

through REST-based Web services. As can be seen in Fig. 1, the general platform is divided into two main modules: a mobile application and the module of intelligent services for health self-management. The mobile application consists in a set of graphical interfaces through which patients provide information concerning health parameters (e.g. weight, blood pressure, heart rate, glucose, and cholesterol), daily physical activities, food intakes, insulin intakes as well as their mood. The second module covers all those functionalities related to the health self-management such as health parameters monitoring, physical activity, food, medication and mood monitoring, alerts management, generation of health recommendations and follow-up of treatments. Furthermore, all this information can be consulted by health professionals to know the current health status of their patients. Next sections provide a description in detail of all modules mentioned above as well as their interrelationships.

3.1 Parameters Monitoring

The design and development of Health Monitor involved health professionals with a wide experience on the treatment of people with diabetes mellitus and hypertension. The involvement of these professionals aimed to effectively determine all those parameters that must be collected and measured along the treatments corresponding to the different chronic diseases to which this platform is focused on. At the end of an iterative process, the health parameters that were established to be monitored by the platform are:

- Blood pressure. This health parameter refers to the strength of the blood pushing against the sides of the blood vessels. Hypertension (high blood pressure) does not tend to produce obvious symptoms. Therefore, it is important to constantly measure the blood pressure. The blood pressure measurement considers how quickly blood is passing through the veins as well as the amount of resistance the blood meets while it's pumping. Figure 2c shows the mobile interface for registering the blood pressure, whose readings have two values, for example 140/90 mmHg. The first value represents the systolic blood pressure, i.e. the highest pressure when the heart beats and pushes the blood round the body. The second value is the diastolic blood pressure, i.e. the lowest pressure when the heart relaxes between beats.
- Body Mass Index (BMI). This parameter determines the level of body fat. It can help health professionals to determine the overall fitness of patients and their risk of developing chronic diseases or health complications. BMI is important because, on the one hand, a low BMI index can be a signal of malnourished which can be due to the body isn't properly absorbing nutrients. On the other hand, having a high BMI index alerts health professional that their patients' risks of heart disease and diabetes are higher than someone with a normal BMI index.

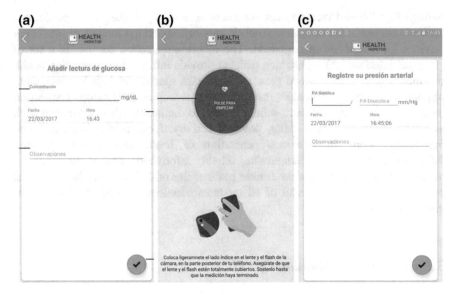

Fig. 2 Health parameters monitoring

- Blood glucose levels. Blood glucose tests measure the amount of glucose in the blood. Out of control blood glucose levels can lead to serious term problems such as hypoglycemia and hyperglycemia. Also, in the long run, uncontrolled blood sugar can damage the vessels that supply blood to organs. Due to these reasons, it is important to monitor this parameter to take action as soon as possible. Figure 2a depicts the mobile interface for registering blood glucose levels. As can be seen, in addition to the blood glucose level, the patient can register observations about the values provided.
- Pulse. Pulse, also known as heart rate, refers to number of times the heart beats per minute. It can help to monitor the fitness level of people. The pulse should be steady; however, it is normal that it varies in response to exercise, excitement, fear, etc. When the heart is beat out of rhythm (either too fast or too slow) patient must discuss this with the health professional. Figure 2b shows the mobile interface for registering the pulse. The pulse can be manually registered; however, in Fig. 2b, this parameter is obtained through the heart rate monitor included in some smartphones.

On the other hand, there are many factors that affect how well chronic diseases such as diabetes mellitus and hypertension are controlled. Fortunately, most of these factors can be controlled by the patients with above-mentioned diseases. For instance, they can choose how much and what they eat, how much exercise they do, how frequently they monitor the blood sugar, as well as the medication dosing [34]. Therefore, to effectively manage diabetes mellitus and hypertension, it is necessary to balance dietary, the physical activity performed as well as the medication. The improvement of patients' lifestyle represents one of the most important factors in the management of chronic diseases. Also, has been proved that an efficient lifestyle produces effective effects on weight and cardiovascular risk factors [35].

Considering the facts above discussed, this module allows patients to register their physical activity, food, medication and mood. All these aspects are important during the chronic disease treatment since they can determine the fail or success of the treatment. Figure 3a shows the mobile interface that allows patients to register their mood. For instance, the user can select the mood among a set of options such as calm, content, bad, indifferent, among others. Meanwhile, Fig. 3b represents the mobile application for the register of insulin intakes. It must be remarked that this is only an example of the type of medication that can be registered in the application. Some examples of medication are prandial glucose regulators, sulphonylureas, metformin, to mention but a few.

Finally, Fig. 3c shows the interface that allow patients to track their meals. The information collected in this mobile interface is concerning how many calories or grams of protein, carbs and fat are in the food. All these data are important for diabetes treatment because glucose comes from carbohydrates foods and normally glucose levels increase slightly after eating.

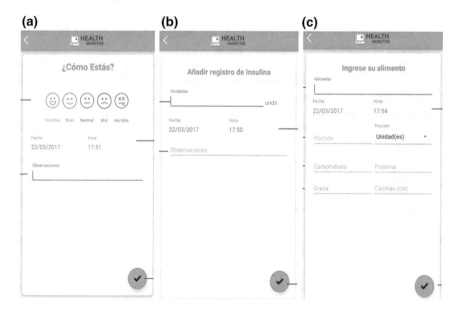

Fig. 3 Physical activity, food, medication and mood monitoring

3.2 Alerts Management

The main goal of this module is to generate a set of alerts when certain risk factors are detected by the platform. For instance, in the context of diabetes management, different alerts are generated when blood glucose levels of the patient are elevated, when the patient consume food with high amounts of fat, carbs or calories, or when the patients forgot to measure their blood glucose levels. As was previously mentioned, a group of healthcare professionals was involved in the design and development of Health Monitor, therefore, they established the set of alerts to be provided to the patients. Also, it must be mentioned that this set of alerts can increase during the deployment of the platform. The alerts generated by this module are related to the next health parameters and food and medication intakes:

- Blood glucose levels. The aim of diabetes treatment is to bring blood sugar levels as close to normal as possible. Therefore, it is necessary that patients know the normal blood sugar level and be alerted when their glucose levels be lower or higher than normal.
- BMI. Being overweight raises the risk for type 2 diabetes mellitus and hypertension. Also, it can increase risk of high blood pressure and unhealthy cholesterol. BMI is interpreted differently for children, teens and adults, as well as among men and woman. Therefore, the platform considers this fact at the moment of generate the respective alerts.
- Cholesterol. The cholesterol is required by the body to function properly; however, having too much cholesterol puts patients at risk for having a heart attack. The total cholesterol consists of low-density lipoproteins (LDL) and high-density lipoproteins (HDL). LDL is called bad cholesterol and HDL is considered good cholesterol. The platform considers factors such as age and gender when alerts about cholesterol are generated.
- Visceral fat. This parameter refers to the body fat that is stored within the abdominal cavity. This type of fat plays a dangerous role that affects hormones function. Storing higher amounts of visceral fat increases the risks of health problems including type 2 diabetes and heart disease. In this sense, the platform generates alerts when high values of visceral fat are provided by the users.
- Blood pressure. When the blood pressure is too high, it puts extra strains on the arteries and heart, which may lead to heart attacks and strokes. Therefore, it is important to keep the blood pressure as low as possible. To help patients to treat hypertension, the platform generates a set of alerts when the data about blood pressure provided by the patient is higher than normal.
- Food. Eating healthful meals is an essential part of managing diabetes mellitus, and hypertension. In this sense, the platform generates alerts when the patient eats food with high amounts of fat, carbs and calories that may increase the risk of cholesterol, heart disease, high blood pressure, uncontrolled blood sugar as well as weight gain.
- Medication. Health professionals may prescribe medications depending on patients' health needs. For instance, a treatment for diabetes mellitus may

include insulin, oral medication or a combination of them. Sometimes, patients may require a treatment with multiple drugs if they have additional cardiovascular risk factor. To avoid health complications, it is very important that patients adhere to their medication plan. Therefore, this platform generates a set or alerts and reminders when patients forget to take their respective medication.

Table 1 provides a summary of the most representative ranges of measurement values for blood glucose levels, IMC, cholesterol and blood pressure. Most of the values presented in Table 1 were established by the WHO (World Health Organization). These values are used by this module to provide patients with different alerts that alert them when the data provided to the platform are lower or higher than normal values. It must be remarked that the ranges shown in Table 1 represent the ranges used for adults. Some of these ranges are different for children and teenagers. Health Monitor considers the age range of patients when alerts are provided.

Figure 4 presents the user mobile interfaces related to the alerts management module. Figure 4a shows the alert generated when the blood pressure is lower than normal. More specifically, the alert generated is "His/Her blood pressure is low, it can represent hypotension". Figure 4b present an alert generated when the platform detects that patients eat food with high amounts of fat, carbs and calories. Finally, Fig. 4c shows the mobile interface with a summary of the alerts provided to the patients by the Health Monitor platform.

Table 1 Measurement values for blood glucose, blood pressure, cholesterol and BMI

Parameter	Category	Values
Blood glucose	Pre-prandial	72–110 mg/dl (or 4–7 mmol/l)
	Postprandial	<180 mg/dl (or 10 mmol/l)
IMC	Underweight	Below 18.5
	Normal weight	18.5–24.9
	Pre-obesity	25.0–29.9
	Obesity class I	30.0–34.9
	Obesity class II	35.0–39.9
	Obesity class III	Above 40
Cholesterol	Desirable	Less than 200 mg/dL
	Borderline high	200–239 mg/dL
	High	240 mg/dL and above
Blood pressure	Normal	120/80
	Prehypertension	120–139/80–89
	Stage 1 hypertension	140–159/90–99
	Stage 2 hypertension	160/100

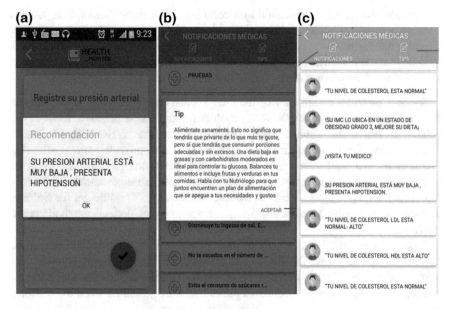

Fig. 4 Examples of alerts generated by Health Monitor

3.3 Health Recommender Module

Recommender systems represent a great technological tool for supporting the self-monitoring of chronic diseases such as diabetes mellitus and hypertension. In this sense, the Health Monitor platform integrates a recommender module that provides patients a set of health recommendations about diets and exercise routines that promote the patient self-management. The recommender module is based on the CF (Collaborative Filtering) recommendation algorithm, which bases the recommendations on the ratings of the user in the system [36].

The health recommender module recommends to the active patient all those items (diets and exercise routines) that other patients with similar tastes liked in the past. For this purpose, this module collects the patients' feedback in the form of rating for diets and exercise routines. Furthermore, this module exploit differences among patients profiles in order to determine how to recommend an item [37]. The patient's profile contains information such as age, weight, chronic disease and information related to the treatment he/she follows including diet, exercise routines, and medication. Figure 5a shows a mobile interface with a diet that has been rated by a patient. Meanwhile, Fig. 5b, c show a set of diets and exercise routines contained in the platform, which are recommended to the patients.

As was previously mentioned, the health recommender module implements the CF recommendation algorithm. More specifically, the MSD (Mean Squared Difference) is used to measure the mean squared difference among two vectors. The MSD's equation is shown below.

(a)   **(b)**   **(c)**

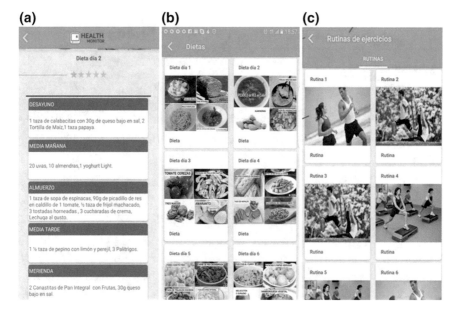

Fig. 5 Health care recommender

$$sim(x, y) = 1 - \frac{1}{\#B_{x,y}} \sum_{i \in l_u} \left(\frac{r_{x,i} - r_{y,i}}{max - min} \right)^2 \in [0, 1] \qquad (1)$$

where $\#B_{x,y}$ represents the number of items that two patients have rated; and $r_{x,i}$ and $r_{y,i}$ represent the rates that were assigned by patient x and patient y respectively. In short, the health recommender module works as follows. Given the patient x and an item i (diet or exercise routine) that has not been evaluated by the patient x, the module determines the group of patients who likes the same items that patient x and that have rated the item i. Based in this information, the module predicts how patient x would rate items that have no evaluated. Finally, the module recommends the n items that were rated with the highest values [38].

3.4 Monitoring of Patients

The lifestyle management of patients with a chronic disease is a difficult task in primary care and traditional face-to-face medical solutions. Health professionals that provide care for patients with diabetes mellitus or hypertension must be informed about the current health status of their patients. Information such as the monitored by this platform can help health care professionals to understand the part

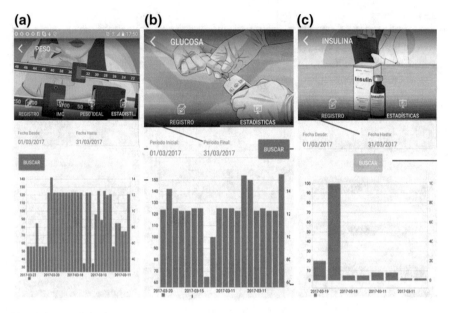

Fig. 6 Monitoring of patients

of the treatment that affects or improve the health status of their patients and then establish a plan of action to improve the quality of life of their patients.

To address the above-discussed fact, the mobile application provides a set of user interfaces through which health professionals can visualize a summary of the data about their patients' health parameters such as blood pressure, blood glucose levels, insulin intakes, calories, carbs and fat grams consumed, calories burned during exercises, among others. All information above mentioned is provided by means of a set of mobile interfaces. For instance, Fig. 6a presents the history of the patients' weight. Meanwhile, Fig. 6b, c presents the history of glucose levels and insulin intakes respectively. Furthermore, health professionals can select a specific period to be more specific about how much time patients have presented health problems.

4 Evaluation and Results

Despite Health Monitor is composed of different modules, the evaluation performed in this work aims to evaluate the effectiveness of the health recommender module i.e. to determine how effective is Health Monitor to provide health recommendations to a specific patient considering the similitude among he/she and the rest of patients registered in the system. The evaluation process performed in this work is composed of three main steps namely selection of participants, generation of recommendations and analysis of results. Next sections provide a wide description of these steps.

4.1 Selection of Participants

The evaluation presented in this work was performed at the Valdivia IESS (Ecuadorian Social Security Institute) Ambulatory Hospital. The evaluation process involved a total of 20 patients and 8 healthcare professionals. On the one hand, patients were eligible for inclusion in this evaluation if they meet the next criteria: (1) adults between 18 and 50 years of age; (2) patients with Type 2 diabetes mellitus or hypertension; (3) patients can use mobile devices as well as mobile applications. These inclusion criteria are aligned with the main objective of this evaluation, which is to determine whether the recommender system works under conditions that will be usual for the Health Monitor platform. It is important to mention that patients with an insulin pump and those ones on dialysis were excluded from this process since these conditions require specific self-management functionalities that are not currently supported by the platform. Based on the inclusion and exclusion criteria, a total of 10 patients with Type 2 diabetes mellitus and 10 patients with hypertension were selected. On the other hand, the unique criterion that health professionals involved in this process must meet is to have more than one year of experience in the treatment of patients with Type 2 diabetes mellitus or hypertension. In this way, 2 nurses, 2 dieticians and 4 doctors with experience in diabetes and hypertension care were selected.

4.2 Generation of Recommendations

Once the group of patients and healthcare professionals was selected, the mobile application provided by the Health Monitor platform was introduced to them. All functionalities provided by this application were described to the group. Then, the patients were asked to install the application on their mobile device (either smartphone or tablet) as well as to perform the registration process. This process consists in provide information concerning their profile such as age, weight, height and chronic disease.

All patients used the mobile application along three months. During this time, patients rated different diets and exercise routines previously established by healthcare professionals. At the end of the mentioned period, the Health Monitor platform provided, through its health recommender module, a set of recommendations that could be useful for patients to improve their health status. Finally, aiming to evaluate the correctness of the recommendations generated by the platform, the precision, recall and F-measure metrics were used [39]. Despite the fact that these metrics are commonly used to evaluate information retrieval systems [40], recently they also have been used for evaluating recommender systems since these ones are usually considered as personalized information retrieval systems [41]. The formulas of these metrics are presented below.

$$precision = \frac{correctly\ recommended\ items}{total\ recommended\ items} \tag{2}$$

$$recall = \frac{correctly\ recommended\ items}{relevant\ items} \tag{3}$$

$$F\text{-measure} = 2 * \frac{precision * recall}{precision + recall} \tag{4}$$

For the purposes of this evaluation, the precision metric represents the ability of the Health Monitor platform to recommend as many relevant items as possible. Hence, the precision score is obtained by dividing the *correctly recommended items*, which represent the items provided by the system that are classified as relevant by the patient, among the *total recommended items*, which represent the total number of items provided by the system. With respect to the recall score, it represents the ability of the platform to recommend as few non-relevant items as possible. Therefore, this score is obtained by dividing the *correctly recommended items* among the items that are relevant for the patient (*relevant items*). Finally, the F-measure score represents the harmonic mean of its precision and recall [42].

4.3 Analysis of Results

Table 2 presents the results obtained by the health recommender module concerning patients with diabetes mellitus. As can be observed in Table 2, a total of 78 items were recommended by the module. Also, 62 of the 78 items were correctly recommended, which represents a precision of 0.793. In general terms, the health recommender module obtained good results with an average F-measure of 0.812, a precision of 0.793 and a recall of 0.836 for patients with diabetes mellitus. Particularly, the patient with higher precision, recall and F-measure scores is the patient 4, who received as recommendation 8 correct items out of a total of 9.

Regarding patients with hypertension, the evaluation results are shown in Table 3, from which it can appreciated that the health recommender module got similar results that the obtained for patients with diabetes mellitus. The recommender module recommended a total of 74 items from which 59 items were correctly recommended. With regards to the precision, recall and F-measure values, the module obtained the scores of 0.803, 0.830 and 0.816 respectively. In this case, the

Table 2 Evaluation results for patients with diabetes mellitus

Patient	Total	Relevant	Correct	Precision	Recall	F-measure
Patient 1	7	7	6	0.857	0.857	0.857
Patient 2	7	7	5	0.714	0.714	0.714
Patient 3	8	8	7	0.875	0.875	0.875
Patient 4	9	9	8	0.889	0.889	0.889
Patient 5	6	5	4	0.667	0.800	0.727
Patient 6	7	7	6	0.857	0.857	0.857
Patient 7	10	8	7	0.700	0.875	0.778
Patient 8	7	6	5	0.714	0.833	0.769
Patient 9	10	10	8	0.800	0.800	0.800
Patient 10	7	7	6	0.857	0.857	0.857
	78	74	62	0.793	0.836	0.812

Table 3 Evaluation results for patients with hypertension

Patient	Total	Relevant	Correct	Precision	Recall	F-measure
Patient 1	6	6	5	0.833	0.833	0.833
Patient 2	10	9	7	0.700	0.778	0.737
Patient 3	5	5	4	0.800	0.800	0.800
Patient 4	8	8	7	0.875	0.875	0.875
Patient 5	8	7	6	0.750	0.857	0.800
Patient 6	5	5	4	0.800	0.800	0.800
Patient 7	6	6	5	0.833	0.833	0.833
Patient 8	10	9	8	0.800	0.889	0.842
Patient 9	9	9	7	0.778	0.778	0.778
Patient 10	7	7	6	0.857	0.857	0.857
	74	71	59	0.803	0.830	0.816

patient that obtained the highest scores was the number 4, with a precision, recall and F-measure of 0.875.

In this evaluation, we found that the health recommender module got encouraging results regardless of the disease to which it was focused on. Considering the results obtained, we believe that Health Monitor can improve patient knowledge at little cost through the recommendation of diet and exercise routines endorsed by healthcare professionals with a wide experience in the care of chronic disease patients. This fact may allow patients to improve their lifestyle through the correct control and management of their disease. Furthermore, it is expected that Health Monitor can contribute to decreasing the medical costs needed to the attention and treatment of patients with this kind of diseases. This contribution will be possible in part thanks to the growing number of patients using mobile devices as well as the growing use of health and fitness mobile applications. Finally, it must be clear that the platform presented in this work is designed to support primary care providers, and it is not intended to be a total substitution of the patient-healthcare professional relationship.

5   Conclusions

This chapter presented the general architecture of Health Monitor, a platform for the self-management of the chronic diseases diabetes mellitus and hypertension. The main goal of Health Monitor is to empower patients to be more active in managing their chronic disease as well as to contribute to the reduction of the economic costs related to the management and treatment of chronic diseases. However, despite the use of mobile applications for chronic disease self-management is promising, there are several issues that must be addressed to achieve the above-mentioned goals. Some issues that were identified during the development of the Health Monitor platform are: (1) the necessity of keeping patients' motivation high i.e., to keep the interest of patients for their disease management. In this sense, it is necessary to establish new mechanisms that make disease management an easy and funny task; (2) to provide alternative support methods for children and older adults with a chronic disease in order to allow them to effectively and independently use mobile applications for the management and control of chronic diseases; (3) to promote the adoption of mobile technologies for the treatment of chronic disease as well as the continuous participation of patients in health self-management, and (4) to provide secure and transparent mobile applications that preserve the privacy of the patients.

As future work, we plan to make the platform as autonomous as possible, i.e., we plan to reduce the user inputs as much as possible. For this purpose, we will analyze market-available wearables that automatically monitor vital parameters such as glucose level, blood pressure, and weight. In this sense, we will analyze the investment cost that the use of wearables represents to make this platform accessible to a wide range of people of different economic levels. On the other hand, we plan to integrate mechanisms that allow health professionals that provide care for patients with chronic diseases to analyze and learn diseases treatments that have been proved to be effective with members of the platform and then use them with similar patients. Furthermore, the next phase of this project will gather requirements and attitudes from doctors and patients of different chronic diseases such as obesity to allow Health Monitor to provide support to these diseases.

Acknowledgements This work has been funded by the Universidad de Guayaquil (Ecuador) through the project entitled "Tecnologías inteligentes para la autogestión de la salud". Finally, this work has been also partially supported by the Spanish National Research Agency (AEI) and the European Development Fund (FEDER /ERDF) through project KBS4FIA (TIN2016-76323-R).

References

1. Bährer-Kohler, S., Krebs-Roubicek, E.: Chronic disease and self-management—aspects of cost efficiency and current policies. In: Self Management of Chronic Disease, pp. 1–13. Springer, Berlin (2009)
2. WHO | World Health Organization: Global report on diabetes. World Health Organization (2016)
3. Ren, L., Han, W., Yang, H., Sun, F., Xu, S., Hu, S., Zhang, M., He, X., Hua, J., Peng, S.: Autophagy stimulated proliferation of porcine PSCs might be regulated by the canonical Wnt signaling pathway. Biochem. Biophys. Res. Commun. **479**, 537–543 (2016)
4. OECD.: OECD Reviews of Health Care Quality: Korea 2012. OECD Publishing, Paris (2012)
5. Wager, K.A., Lee, F.W., Glaser, J.P. (John P.: Health care information systems: a practical approach for health care management (2013)
6. Zwar, N., Harris, M., Griffiths, R., Roland, M., Dennis, S., Powell, G., Hasan, I.: A systematic review of chronic disease management, Australia (2006)
7. Zheng, Z., Ma, H., Lyu, M.R., King, I.: WSRec: A collaborative filtering based web service recommender system. In: 2009 IEEE International Conference on Web Services, pp. 437–444. IEEE (2009)
8. Tejeda-Lorente, Á., Porcel, C., Peis, E., Sanz, R., Herrera-Viedma, E.: A quality based recommender system to disseminate information in a university digital library. Inf. Sci. (Ny) **261**, 52–69 (2014)
9. Yuan, N.J., Zheng, Y., Zhang, L., Xie, X.: T-Finder: a recommender system for finding passengers and vacant Taxis. IEEE Trans. Knowl. Data Eng. **25**, 2390–2403 (2013)
10. Colombo-Mendoza, L.O., Valencia-García, R., Rodríguez-González, A., Alor-Hernández, G., Samper-Zapater, J.J.: RecomMetz: a context-aware knowledge-based mobile recommender system for movie showtimes. Expert Syst. Appl. **42**, 1202–1222 (2015)
11. Martínez-Pérez, B., de la Torre-Díez, I., Candelas-Plasencia, S., López-Coronado, M.: Development and evaluation of tools for measuring the quality of experience (QoE) in mHealth Applications. J. Med. Syst. **37**, 9976 (2013)
12. Ventola, C.L.: Mobile devices and apps for health care professionals: uses and benefits. Pharm. Ther. **39**, 356–364 (2014)
13. Andrus, M.R., Forrester, J.B., Germain, K.E., Eiland, L.S.: Accuracy of pharmacy benefit manager medication formularies in an electronic health record system and the Epocrates mobile application. J. Manag. Care Spec. Pharm. **21**, 281–286 (2015)
14. Doukas, C., Pliakas, T., Maglogiannis, I.: Mobile healthcare information management utilizing Cloud Computing and Android OS. In: 2010 Annual International Conference of the IEEE Engineering in Medicine and Biology, pp. 1037–1040. IEEE (2010)
15. Abbasgholizadeh Rahimi, S., Menear, M., Robitaille, H., Légaré, F.: Are mobile health applications useful for supporting shared decision making in diagnostic and treatment decisions? Glob. Health Action **10**, 1332259 (2017)
16. Fruehauf, H., Knobloch, N., Knobloch, S., Semela, D., Vavricka, S.R.: The "HCV Advisor" App—a web-based mobile application to identify suitable treatments with direct antiviral agents for chronic hepatitis C infection. J. Hepatol. **66**, S507 (2017)
17. Wallace, S., Clark, M., White, J.: "It's on my iPhone': attitudes to the use of mobile computing devices in medical education, a mixed-methods study. BMJ Open **2**, e001099 (2012)
18. Ozdalga, E., Ozdalga, A., Ahuja, N.: The smartphone in medicine: a review of current and potential use among physicians and students. J. Med. Internet Res. **14**, e128 (2012)
19. Chang, S.-H., Li, C.-N.: A cloud based type-2 diabetes mellitus lifestyle self-management system. In: Trends and Applications in Knowledge Discovery and Data Mining. **9441**, 91–103 (2015)
20. Desveaux, L., Agarwal, P., Shaw, J., Hensel, J.M., Mukerji, G., Onabajo, N., Marani, H., Jamieson, T., Bhattacharyya, O., Martin, D., Mamdani, M., Jeffs, L., Wodchis, W.P.,

Ivers, N.M., Bhatia, R.S.: A randomized wait-list control trial to evaluate the impact of a mobile application to improve self-management of individuals with type 2 diabetes: a study protocol. BMC Med. Inform. Decis. Making **16**, 144 (2016)

21. Nguyen, H.D., Jiang, X., Poo, D.C.C.: Designing a social mobile platform for diabetes self-management: a theory-driven Perspective. Social Computing and Social Media. **9182**, 67–77 (2015)

22. Sebillo, M., Tucci, M., Tortora, G., Vitiello, G., Ginige, A.: M-Health and self care management in chronic diseases—territorial intelligence can make the difference. Digitally Supported Innovation. **18**, 207–219 (2016)

23. Sunyaev, A., Chornyi, D.: Development of an Internet-based chronic disease self-management system. Pervasive Health Knowledge Management. Healthcare Delivery in the Information Age. **9194**, 271–283 (2013)

24. Whitlock, L.A., McLaughlin, A.C., Harris, M., Bradshaw, J.: The design of mobile technology to support diabetes self-management in older adults. Human Aspects of IT for the Aged Population. **9194**, 211–221 (2015)

25. Demidowich, A.P., Lu, K., Tamler, R., Bloomgarden, Z.: An evaluation of diabetes self-management applications for Android smartphones. J. Telemed. Telecare. **18**, 235–238 (2012)

26. Petersen, M., Hempler, N.F.: Development and testing of a mobile application to support diabetes self-management for people with newly diagnosed type 2 diabetes: a design thinking case study. BMC Med. Inform. Decis. Mak. **17**, 91 (2017)

27. Hardinge, M., Rutter, H., Velardo, C., Shah, S.A., Williams, V., Tarassenko, L., Farmer, A.: Using a mobile health application to support self-management in chronic obstructive pulmonary disease: a six-month cohort study. BMC Med. Inform. Decis. Making **15**, 46 (2015)

28. Ariaeinejad, M., Archer, N., Stacey, M., Rapanos, T., Elias, F., Naji, F.: User-centered requirements analysis and design solutions for chronic disease self-management. Presented at the July 17 (2016)

29. Sookpalng, C., Vanijja, V.: Design of disease management system for chronic heart failure: a case study from advanced heart failure clinic at King Chulalongkorn Memorial Hospital. Presented at the December 12 (2013)

30. Park, H.S., Cho, H., Kim, H.S.: Development of a multi-agent m-health application based on various protocols for chronic disease self-management. J. Med. Syst. **40**, 36 (2016)

31. Jung, H., Chung, K.: Life style improvement mobile service for high risk chronic disease based on PHR platform. Cluster Comput. **19**, 967–977 (2016)

32. Leijdekkers, P., Gay, V.: Mobile apps for chronic disease management: lessons learned from myFitnessCompanion®. Health Technol. (Berl) **3**, 111–118 (2013)

33. Schwaber, K., Beedle, M.: Agile Software Development with Scrum. Prentice Hall (2002)

34. Tama, B.A., Rodiyatul F. S., R.F.S., Hermansyah, H.: An early detection method of type-2 diabetes mellitus in Public Hospital. TELKOMNIKA (Telecommunication Comput. Electron. Control. **9**, 287 (2011)

35. Danaei, G., Finucane, M.M., Lu, Y., Singh, G.M., Cowan, M.J., Paciorek, C.J., Lin, J.K., Farzadfar, F., Khang, Y.-H., Stevens, G.A., Rao, M., Ali, M.K., Riley, L.M., Robinson, C.A., Ezzati, M.: Global Burden of Metabolic Risk Factors of Chronic Diseases Collaborating Group (Blood Glucose): National, regional, and global trends in fasting plasma glucose and diabetes prevalence since 1980: systematic analysis of health examination surveys and epidemiological studies with 370 country-years and 2.7 million participants. Lancet **378**, 31–40 (2011)

36. Ekstrand, M.D., Riedl, J.T., Konstan, J.A.: Collaborative filtering recommender systems. Found. Trends Hum. Comput. Interact. **4**, 81–173 (2011)

37. Melville, P., Mooney, R.J., Nagarajan, R.: Content-boosted collaborative filtering for improved recommendations. In: Eighteenth National Conference on Artificial Intelligence, 1034 (2002)

38. Shardanand, U., Maes, P.: Social information filtering: algorithms for automating "word of mouth." In: Proceedings of the SIGCHI Conference on Human Factors in Computing Systems —CHI '95, pp. 210–217. ACM Press, New York (1995)

39. Salton, G., McGill, M.J.: Introduction to Modern Information Retrieval. McGraw-Hill (1983)

40. Raghavan, V., Bollmann, P., Jung, G.S.: A critical investigation of recall and precision as measures of retrieval system performance. ACM Trans. Inf. Syst. **7**, 205–229 (1989)

41. Fang, B., Liao, S., Xu, K., Cheng, H., Zhu, C., Chen, H.: A novel mobile recommender system for indoor shopping. Expert Syst. Appl. **39**, 11992–12000 (2012)

42. Dumais, S., Balazinska, M., Hwang, J.-H., Shah, M.A., Schettini, R., Ciocca, G., Gagliardi, I., Dash, M., Koot, P.W., Bustos, B., Schreck, T., Plachouras, V., Goodchild, M.F., Tannen, V., Tannen, V., Jensen, C.S., Snodgrass, R.T., Zhang, A., Bhargava, B., Gibbons, P.B., Zhang, E., Zhang, Y., Chakrabarti, S., Deutsch, A., Jensen, C.S., Snodgrass, R.T., Kennedy, J., Cannon, A., Arenas, M., Gray, P.M.D., Deng, K., Woodruff, D., Huan, J., Metwally, A., Leung, C.K.-S., Cheng, H., Han, J., Ukkonen, A., Ziegler, C.-N., Gray, P.M.D., Dobbie, G., Ling, T.W., Kolahi, S., Gray, P.M.D., Pasi, G., Novák, V., Novák, V., Novák, V., Novák, V.: F-measure. In: Encyclopedia of Database Systems, pp. 1147–1147. Springer US, Boston (2009)

Reliable and Smart Decision Support System for Emergency Management Based on Crowdsourcing Information

Karina Villela, Claudia Nass, Renato Novais, Paulo Simões Jr.,
Agma Traina, Jose Rodrigues Jr., Jose Manuel Menendez,
Jorge Kurano, Tobias Franke and Andreas Poxrucker

Abstract Command and control centres face the challenge of quickly obtaining accurate information about emergencies they should response to. Conversely, crowdsourcing information and mobile technologies offer great potential for better engaging eyewitnesses in emergency and crisis management processes. This paper describes the vision and the realisation of the RESCUER system, a smart and interoperable decision support system for emergency and crisis management based on mobile crowdsourcing information. Eight evaluation exercises with end users were performed during the project duration, in addition to technical verifications of the individual system components. The results of the evaluation exercises were quite positive and helped to continuously improve and extend the system.

Keywords Emergency management · Crowdsourcing information
Multimedia data analysis · Decision support system

K. Villela (✉) · C. Nass
Fraunhofer Institute for Experimental Software Engineering, Kaiserslautern, Germany
e-mail: Karina.Villela@iese.fraunhofer.de

R. Novais
Fraunhofer Project Center for Software and Systems Engineering,
Federal University of Bahia, Instituto Federal de Educacão,
Ciência e Tecnologia da Bahia, Salvador, Brazil

P. Simões Jr.
Fraunhofer Project Center for Software and Systems Engineering,
Federal University of Bahia, Salvador, Brazil

A. Traina · J. Rodrigues Jr.
University of Sao Paulo, Sao Carlos, Brazil

J. M. Menendez · J. Kurano
Polytechnic University of Madrid, Madrid, Spain

T. Franke · A. Poxrucker
German Research Center for Artificial Intelligence (DFKI), Kaiserslautern, Germany

© Springer International Publishing AG 2018
R. Valencia-García et al. (eds.), *Exploring Intelligent Decision Support Systems*,
Studies in Computational Intelligence 764,
https://doi.org/10.1007/978-3-319-74002-7_9

1 Introduction

Emergencies are defined as critical situations caused by incidents, natural or man-made, that require measures to be taken immediately to reduce their adverse consequences to life and property [1, 2]. Examples of natural incidents are floods, wildfires and snow storms, whereas examples of man-made incidents are fires, explosions and substance spills from oil platforms, ships and factories. These and other man-made incidents can also occur during large-scale events, either by negligence or on purpose, as for instance in the case of a terrorist attack. In large-scale events, the risk of crowd-related incidents, such as conflicts between opposing crowds in football matches or mass panic, is of particular concern. The adverse and therefore undesired consequences of an emergency give rise to a crisis.

In this context, emergency management is the planning and establishment of organisational and procedural conditions to control a critical situation and quickly restore normality [3]. Crisis management involves: (1) evaluating the severity of possible adverse consequences of a critical situation; (2) coordinating the required measures for avoiding, controlling and/or mitigating those adverse consequences; and (3) establishing a strategy for communication with the general public and the parties involved in the crisis. In these processes, the main challenge is to quickly obtain contextual information about the situation in order to make the right decisions. According to Engelbrecht et al. [4], late decisions or decisions based on inaccurate information have great potential for causing further damages to life and property.

In a common scenario, an eyewitness speaks on the phone to someone from an operational force and informs him/her about the situation. The information is very often inaccurate and incomplete. Therefore, the command centre has to take generic and standard measures and wait until someone from the operational forces (i.e., a formal responder) arrives at the incident site and can provide more accurate and complete information. Furthermore, getting information later on about the evolution of the situation is also a big challenge for the command centre: Time pressure, resource availability, heterogeneous information and cognitive overload result in an irregular flow of information.

Crowdsourcing information and mobile technologies offer great potential for better engaging eyewitnesses in the emergency and crisis management processes. With the appropriate means, eyewitnesses can send reports of an incident as soon as it occurs or even when realising its imminent risk of occurrence. For example, a YouTube video [5] lasting about 47 min exists, containing excerpts of videos recorded by eyewitnesses of the crowd crush incident that occurred at the Love Parade that took place in Duisburg, Germany, on July 24th, 2010. The incident resulted in 21 people being crushed to death and 500 others being injured. Instead of being used to document the tragedy afterwards, those videos could have assisted in detecting the incident earlier and in monitoring it continuously. However, two issues have to be addressed in order to effectively and efficiently benefit from incident reports coming from eyewitnesses: (1) the reliability of the incident reports, as eyewitnesses might, for different reasons, provide wrong assessments, and (2) the potentially high number of incident reports that need to be processed and visualised at the command centre in an appropriate format.

The RESCUER project [6] developed a smart and interoperable decision support system for supporting emergency and crisis management based on crowdsourcing information. The focus of the project was on incidents occurring in industrial areas or at large-scale events, but the resulting platform can be extended to deal with critical situations in other contexts. The project consortium included COFIC, an association encompassing the 90 companies located in the Petrochemical Park of Camacari, Brazil, which is the largest industrial park south of the equator.

The remainder of this chapter reports the vision, implementation and evaluation of RESCUER.

2 Related Work

Several emergency management systems that can be classified as decision support systems have been developed. Many of them (e.g., CRISMA [7], the Rescue Assistant System presented by Technical University of Clausthal [8], CIVIL [9], the Integrated Disaster Management Support System presented by Tomoyuki et al. [10] and PIECES [11]) do not take into consideration the use of crowdsourcing information. On the other hand, Poblet et al. [12] present a review of crowdsourcing tools for emergency management that includes 25 tools (16 emergency management platforms and 9 mobile applications). Several of them are general-purpose tools (but can or have been used for supporting emergency management) and/or cannot be classified as decision support systems. CrisisTracker [13], Sahana EDEN [14] and Ushahidi [15–17] are examples of dedicated emergency management platforms related to RESCUER. CrisisTracker automatically tracks sets of keywords on Twitter and constructs stories by clustering related tweets on the basis of their lexical similarity. It also enables users to verify and analyse stories. Sahana EDEN is a free and open-source disaster management platform that focuses on relief operation, recovery and rehabilitation. It offers organisation registry, volunteer management, crisis mapping and supply management. The most recent development efforts are an alerting and messaging broker and the use of pictographs for disaster communication via mobile applications. Last but not least, Ushahidi is a well-known open-source platform for crisis mapping that has strongly invested in crowdsourcing information. In its early use cases [15, 16], it helped in the planning of relief distribution based on information sent by the affected population via SMS, which needed to be analysed by humans in order to identify information such as geo-location. More recently [17], it provides a mobile application that allows better identification of such information, but still relies on filters and saved searches for access to relevant information.

RESCUER goes beyond the tools reviewed in [12] and establishes the need for the extension of the classification scheme proposed by the authors. In RESCUER, the crowd has the role of sensors and of reporters, but RESCUER is based on the conviction that social media platforms are not appropriate for reporting incidents at the incident site when people are still under stress and therefore provides a dedicated mobile application with three interaction modes to be used according to the stress level of its users (Sect. 4.1). In this scenario, users are not expected to provide any kind of metadata or support tagging. Metadata is collected automatically and tags are extracted by the data analysis components. To this end, RESCUER adds image and video analysis to the core functionalities mentioned in the classification. Of course, the project focused only on a few image or video analysis components (Sects. 4.4 and 4.5). Based on a lesson learned from domain experts in the application domain of emergency management in industrial parks and at large-scale events, RESCUER offers filtering, mapping and navigation functionalities over reports like other pieces of related work do, but focuses on the provision of optimised views of aggregated information on the level of incidents (Sects. 4.6 and 4.7). Finally, RESCUER does not support management of volunteers as Sahana EDEN does, which is one of the core functionalities mentioned in the classification scheme, but was beyond the scope of the project.

3   The RESCUER Vision

Figure 1 illustrates the RESCUER vision. Everything starts with the *Command & Control Staff* setting up the *Emergency Management Session*, which defines the area and the timeframe for emergency management. Several *Crowd Sensing Sessions* can be defined within an *Emergency Management Session* to specify when and where *Movement Data* should be collected from on-site people *(Workforces/ Supporting Forces/Civilians)*. This is particularly relevant in the case of large-scale events that are spread over large areas and last for a longer period of time (e.g., the Olympic Games). A *Crowd Sensing Session* results in the creation of a *Crowd Heat Map*, which uses colours for describing the people density in the area. If no incident is reported, the *Command & Control Staff* simply observes this *Crowd Heat Map*, which is periodically updated to show the current situation.

Whenever an incident takes place, on-site people start sending *Incident Reports*. *Incident Reports* can also come from specific sensor systems installed in the managed area, as they are a class of *Legacy Systems* RESCUER can interface with. *Incident Reports* are visualised by the *Command & Control Staff* in raw format and in aggregated formats (*Incident* Summary and *Emergency Summary*). Based on the available information, the *Command & Control Staff* send *Command & Control Messages* that reflect their decisions on how to handle the reported incidents (and, therefore, control the overall emergency situation) to on-site people. These messages comprise: (1) *Information Requests*, which can be sent to any on-site person, but to civilians only under certain conditions (see Sect. 4.1); (2) *Guidance*

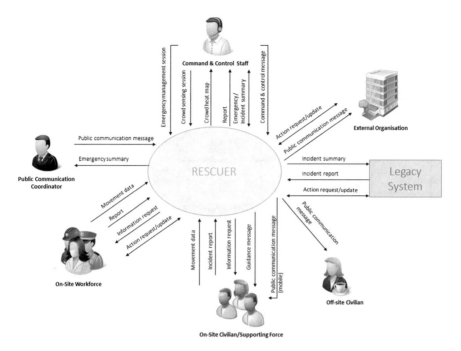

Fig. 1 RESCUER vision

Messages, which are sent to *On-Site Civilians and Supporting Forces;* and (3) *Action Requests*, which can be sent to *On-Site Workforces*, *Legacy Systems* and *External Organisations*. Police, firefighters and rescue services as organisations might have representatives at the *Command & Control Staff* and then *Action Requests* can be directly sent to *On-Site Workforces*. As alternatives, they can communicate with RESCUER through a *Legacy System* (which also receives *Incident Summary* information) or be considered an *External Organisation* (if there is no way of direct communication with RESCUER). Examples of other *External Organisations* are hospitals and military police. In addition, *On-Site Workforces* can send *Query Reports* and *Status Update Reports* to request command centre resources such as health care staff or a helicopter, and inform the arrival of resources and the execution of tasks regardless of whether they are the sender of the query or the person responsible for the task performance.

In order to provide official information about the emergency situation, the *Public Communication Coordinator* receives the *Emergency Summary* and, supported by RESCUER, can quickly send *Public Communication Messages* with customised content and format to several stakeholders, namely *On-Site Civilians*, *Off-Site Civilians* and *External Organisations* (press and public administration offices).

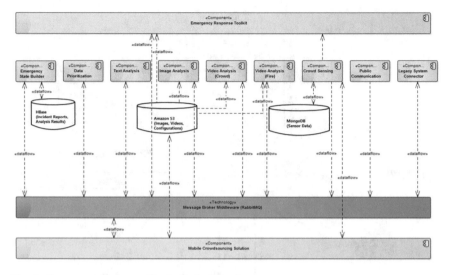

Fig. 2 Component diagram with physical connections

4 The Realisation of the RESCUER Vision

In order to realise the RESCUER vision, the architecture illustrated in Fig. 2 was defined and its components were developed. RESCUER has an event-driven architecture. The components follow the *Microservices architecture* principle [18]. This means that the components are self-sufficient and follow a *shared-nothing* approach, which ensures scalability and performance, as components can run in parallel and independently. To integrate the RESCUER components, the integration platform provides a publish/subscribe mechanism. The components use this mechanism to have asynchronous communication among them. This architecture also allows easy extension of the system, as further multimedia analysis components (microservices) can be included to address aspects other than those currently addressed: fire, smoke and crowd issues. The remainder of this section explains in more detail the purpose and the characteristics of the architectural components.

4.1 Gathering Data

In RESCUER, the data required to support decision-making at the command centre is gathered from the crowd carrying mobile devices at the incident site. In order to make this possible without risking adverse consequences to people's safety, the Mobile Crowdsourcing Solution (MCS) offers two information-gathering mechanisms. In the first one, the Mobile Crowdsourcing Solution collects *Movement Data* (position, direction and speed) according to the defined *Crowd Sensing Sessions* (sensing area, start time, end time) if its user has explicitly consented to the provision of this kind of data to the command centre in order to support the early identification of problems related to crowd behaviour (e.g., high people density, crowd pressure against physical barriers and crowd portions moving in opposite directions). Data collection takes place automatically, meaning without any interaction between users and their devices. In the second information-gathering mechanism, eyewitnesses (civilians and supporting forces working at the venue) open the Mobile Crowdsourcing Solution to report an incident that has just taken place. *Incident Reports* can be quick reports (type of incident, which is the only information provided by the user, plus position and time of the report, which are collected automatically) or provide as much information as the person is capable and willing to provide. The currently covered incident types are: fire, gas leakage, explosion, environmental, vandalism, people crush, mass collapse, stampede and shooting, but the user can also simply indicate "Help". The capability of people to provide further information about the incident depends on their stress level, whereas their willingness depends on their assessment of priorities. According to the information-gathering strategy defined to address both aspects [19], the interaction of users with the Mobile Crowdsourcing Solution can evolve from *One Click Interaction*, which supports quick reports; via *Guide Interaction*, which provides a small set of pre-defined questions with possible answers as well as the possibility of taking photos; to *Chat-like Interaction*, where the users can provide any information they want in free text messages and send as many photos and videos as they want. Furthermore, civilians will only receive *Information Requests* from the command centre if they are in a safe place and have already interacted with the system (which indicates that they are emotionally capable of receiving and answering the *Information Request*).

4.2 Prioritising Data

As communication network traffic increases in most emergencies [20], there might be a huge number of reports coming to the RESCUER backend at the same time. Thus the Data Prioritisation component implements a technique for ensuring that incident reports with the potential for providing new information or confirming unreliable information will receive priority for image/video data analysis. The goal is to allow the command centre to quickly and appropriately respond to the emergency, thereby restoring normality in the area as soon as possible [21]. The RESCUER data prioritisation scheme takes into consideration: (1) whether the currently available image/video analysis components can extract relevant information for the type of incident just being reported or for the types of incident that are active in the *Emergency Management Session*; (2) whether the occurrence of the reported incident has already been confirmed by previous data analysis results or by a report coming from a workforce; (3) whether all incident information that can be confirmed through the currently available image/video data analysis components have already been confirmed by reports coming from workforces or by previous data analysis results; (4) whether the sender's profile is workforce and there is information that was not provided in a structured way in the report, but might be provided in images/videos attached to the report; (5) whether the received report is an answer to an *Information Request* previously sent by the command centre; (6) whether the type of incident is undefined in the report (Help report); and (7) whether the report contains information about injured people that has not been confirmed yet. Based on these criteria, the Data Prioritisation component assigns a priority level from 0 (lowest priority) to 9 (highest priority) to a received report, which is used by the data analysis components to determine which reports to process first.

4.3   Analysing Text

The Text Analysis component gets as input a free text message that might be provided as part of an incident report (together with some metadata such as the time the message was created or the GPS position of the device when the message was sent), applies a series of (pre-)processing steps to extract semantic content and returns a structured representation of the relevant information. This structured representation is composed of: (1) the normalized text; (2) the extracted W^4H (what, where, who, when, how) phrases; and, whenever possible, (3) some incident attributes (incident type, timestamp, location, number of injured people, number of dead people). In the normalized text, known abbreviations and "Internet slangs" are replaced by corresponding correct words to increase readability. Despite the extraction of W^4H phrases being the most basic semantic analysis functionality, it can be very useful if it is used to speed up the recognition of relevant information through special visualisation mechanisms (as in Fig. 7). In addition, they provide the basis for the incident attributes. For example, a "when" phrase like "five minutes ago" can be used to automatically correct the incident time assumed from the timestamp when the report was sent out. In terms of (pre-)processing steps, a *Language Identification* subcomponent determines the language of the free text message. The current possibilities are English and Portuguese. As proposed in [22], the *Text Processing* subcomponent comprises: tokenizer, normalisation and sentence splitter as syntactic pre-processing steps; and number tagger, measurement tagger, POS tagger, chunker and gazeteer lookup as semantic pre-processing steps. In RESCUER, the gazetteer lookup provides the keywords that map (sequences of) words to concepts defined in an ontology model [23]. The *Information Extraction* subcomponent determines semantic annotations and infers incident attributes. The *Emergency State*, which is maintained by the Emergency State Builder, is defined by a set of descriptive attributes like the location of the incident, its starting time, type, the number of injured and dead people, and any further properties that can contribute to a comprehensive overview of the situation. The *Emergency State* influences the text analysis and the text analysis updates the *Emergency State*. For example, if the type of incident is leakage, then the Text Analysis component will look for leaked materials in the "how" phrases of the free text message; later on, the extracted information will be added to the *Emergency State*. The Text Analysis component, as an individual component, was evaluated [23] using a corpora of Twitter messages from the following four incidents: the Boston Brownstones fire on the 26th of March 2014; the Boston marathon bombings on the 15th of April 2013; the shooting in San Bernardino, California, on the 2nd of December 2015; and the shooting at the Pulse nightclub in Orlando, Florida, on the 12th of June 2016. The average processing time for a typical Twitter message (approx. 20 words on average) is about 15 ms. The ability to extract relevant W^4H phrases from the tweets was also evaluated. Good results were obtained for "what", "who", "when" and "how". For "where", the approach using keywords and prepositional phrases resulted in low precision. Reasons for this low precision and corresponding measures are discussed in [23].

4.4 Analysing Images

The Image Analysis component is capable of: (1) identifying fire, (2) identifying smoke with colour discrimination, (3) speeding up smoke identification by using a threshold analysis segmentation and (4) identifying (near) duplicate images regardless of their content. Capabilities (1)–(3) are relevant for the automatic analysis of images related to fire or explosion incidents, whereas capability (4) is relevant for reducing the data processing load and the information to be considered by human specialists. The BoWFire method [24], which was developed in the project, is used for identifying fire in images. The method consists of three steps: (a) colour classification, (b) texture classification and (c) region merge. The first step classifies the pixels' colour as fire or non-fire through a Naive Bayes classifier. The second step extracts Regions of Interest (ROI) from the images using a superpixel[1] algorithm; it then extracts texture features from each region of interest (ROI). To classify the texture features as fire or non-fire regions, BoWFire uses the k-Nearest Neighbours (kNN) classifier. The first two steps occur in parallel to produce images in which fire-classified pixels are annotated. Then, in the third step, the outputs from both classifications are merged into a single output image containing only fire regions. One of the advantages of this method is the capability to dismiss false-positives. In particular, BoWFire allows discarding images with fire-like regions, such as sunsets, and red or yellow objects. The method and its corresponding algorithm were validated [24] using the BoWFire image dataset [25], a dataset composed of 226 images with various resolutions, in which 119 images depict fire and 107 images do not depict fire. The fire images consist of emergencies with different fire incidents, such as buildings on fire, industrial fire, car accidents

---

[1]Region with contiguous pixels of the same colour.

and riots. The remaining images depict emergency situations with no visible fire as well as images with fire-like regions, such as sunsets or red or yellow objects.

Smoke identification is performed using the SmokeBlock method [26], which was also developed in the project and which consists of three steps: (i) First, superpixels are extracted from the input images. Then, in the feature extraction step (ii), information regarding texture and colour is extracted from each superpixel, and one feature vector is generated for each one of them. Finally, in the classification step (iii), each superpixel is classified as "smoke" or "non-smoke" using a Naive Bayes classifier. Based on this information, the SmokeBlock method builds a segmented image, which is composed of the superpixels classified as smoke. The colour of the smoke corresponds to the mean colour of each smoke region. Depending on the information contained in the incident report, the Image Analysis component can select between using: (1) the SmokeBlock algorithm, which looks for the presence of different colours in the smoke regions of interest; or (2) a faster approach for detecting grayscale smoke regions. The faster approach consists of two steps: (i) identifying potential areas covered by smoke using a threshold value, then (ii) removing false-positive pixels by applying further thresholds and spatial analysis. The threshold parameters were chosen and optimised through empirical investigations and analyses. Our approach for smoke identification was validated [26] with a dataset containing 832 images labelled as smoke and 834 images labelled as non-smoke.

In order to identify images that are exactly the same or very similar, RESCUER uses the pHash algorithm [27]. This algorithm first converts the colour image into a grayscale image and then the image resolution is shrunk to 32×32. This procedure removes all high frequencies and unnecessary details. As required by the algorithm, resizing or stretching the image does not affect the computation of its pHash value. By using the Hamming distance over the pHash values, it is possible to infer if one image is similar to another.

4.5 Analysing Videos

Video analysis in RESCUER is performed by two dedicated components: a component for the analysis of fire aspects and a component for the analysis of crowd aspects. The video analysis component for fire aspects implements a method named SPATFIRE [28], which was developed in the RESCUER project. As colour is the most representative feature used in fire and flame detection methods, we used a motion colour-based segmentation to capture the patterns of fire. The common approach for related methods relies on algorithms that perform three steps: (1) selection of candidate regions in frames that have fire-like colour pixels, based only on colour information; (2) motion feature extraction from the candidate regions detected in step 1 on subsequent frames; and (3) classification of the motion feature. We proposed a colour model to perform the spatial segmentation that generates the candidate regions for the motion analysis. Our colour model is based on the Hue

Saturation Value colour space. After the spatial segmentation, we extract sparse and dense motion flows from the regions of interest and from the background regions, respectively. In order to attenuate the problem of the camera motion, we developed a technique for estimating and compensating for the motion flow vectors, named "Block-based motion compensation". The final feature vector is a histogram of oriented flows that is classified by an SVM (Support Vector Machines) classifier to determine the segments of the video that contain fire. Our method was validated [28] using three video datasets. The first dataset consists of 27 videos with resolutions varying from 320×240 to 600×336 pixels, and frame rates varying from 10 to 30 Hz with different encodings. We manually annotated each frame of the videos, which resulted in 83,675 frames labelled as "fire" or "non-fire". The second dataset is composed of videos provided by COFIC. It presents a balanced distribution of videos with resolutions varying from 320×240 to 1920×1080 pixels. Those videos were also manually annotated following the same protocol. The last dataset is the BowFire dataset (see Sect. 4.4).

The video analysis component for crowd aspects focuses on estimating crowd density and identifying potential disturbances in crowd behaviour. The key challenge was to deal with non-static and complex environments. Another aspect to consider was the fact that the videos to be recorded using the Mobile Crowdsourcing Solution will mostly contain lateral and frontal views of crowded people, which will result in an extremely high number of occlusions. Based on the work of Dalal and Triggs [29], this video analysis component uses Histograms of Oriented Gradients (HOG) for features extraction and linear SVMs for classification (person/not person). HOG features are robust features that permit discriminating the human form cleanly, even in cluttered backgrounds under difficult illumination conditions. This component was trained using the INRIA Person Dataset [30] in order to achieve a high level of people detection capability. In the analysis phase, the component extracts frames from the received videos, which are scanned at different scales and locations with a sliding window in order to detect as many people as possible. Spatial filtering is performed to eliminate duplicate detections due to the different scale scanning processes. Since videos will arrive without calibration cues, the area values are extrapolated from human sizes. This is possible because the image area occupied by people offers a clue about the real size of the whole image. The number of people detected in a frame is divided by the extrapolated area value to obtain the crowd density estimation for the frame. The final density estimation is calculated based on the frame densities. A threshold for crowd density is used to indicate a potentially dangerous situation.

The identification of potential crowd disturbances is addressed in RESCUER through an optical flow algorithm called TV-L1 [31, 32], which stands for Total Variation regularisation using the robust L1 norm in the data fidelity term. Optical flow is the apparent visual motion that objects cause as a result of the relative motion between the observer and the scene. TV-L1 is based on the idea of preserving discontinuities in the flow field. It offers robustness against illumination changes, occlusions and noise. It is also an algorithm that offers a good trade-off between precision and computational cost. Most importantly, it can give a good cue about disturbances in a scene, since disturbances will produce fast and unrelated motion vectors. The values of optical flow are analysed to discard those with harmonic values. For example, if there is a general tendency in the motion vectors or if these vectors have the same direction, there is no risk of crowd disturbance. The preliminary evaluations of both subcomponents (estimation of crowd density and identification of potential crowd disturbance) are described in [23].

4.6 Aggregating Data

Reports are used to get data about incidents, but incidents are the system entities that really matter to the command centre. A pre-requisite for incident data aggregation is the assignment of reports coming from the managed area to specific incidents according to the type of incident and the position of the report. Once the report is assigned to an existing incident or generates a new incident in the system, the Emergency State Builder component aggregates the different types of data included in the report in order to generate or update the incident description. Figure 3 illustrates how the structured data of a report is used to update the current incident description. The results of image/video analyses are used to either add or update the respective Data Analysis Solution (DAS) branch in the aggregation tree as a different source of information, which means on the same level as civilians, supporting forces or workforces. This keeps the source of each piece of information clear and easy to be queried. The W^4H phrases collected through the analysis of free text messages are aggregated in a similar way, with the roots of the aggregation trees being the W^4H questions. This aggregation scheme was approved by domain experts, who pointed out that they need to know what is being reported and how reliable the information is, which is extrapolated from the source of the information and the number of reports containing the same information. In particular, workforces are the most reliable source of information, whereas civilians are the least reliable source of information. When the results of image/video analyses

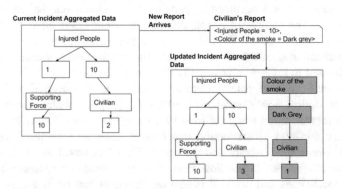

Fig. 3 Aggregation of structured data

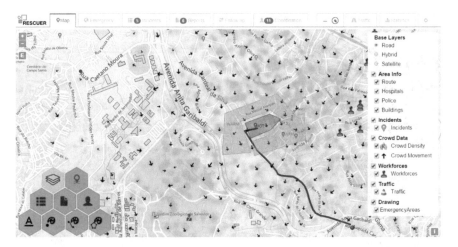

Fig. 4 Visualisation of aggregated movement data including drawn annotations

corroborate structured data, the RESCUER system can link the analysed media to the structured data, which allows the end user to quickly confirm what the crowd is reporting.

Finally, *Movement Data*, which is anonymised anyway, is clustered to indicate crowd movement. The intention is not to present data from individuals, but rather depict the behaviour pattern of portions of the crowd. This clustering allows identifying areas where the level of people density or crowd behaviour is considered dangerous (see Fig. 4).

4.7 Visualising Relevant Data

One of the most important contributions of RESCUER compared to related work is the decision-making support that it provides. Instead of plotting incident reports on a map and letting the *Command & Control Staff* find relevant information in the myriad of incident reports, RESCUER directly presents relevant aggregated information on the incident level, which is only possible due to its data processing capabilities. The screen in Fig. 4 shows on a map the behaviour pattern of portions of the crowd. People density is represented in terms of a heat map; movement direction is represented by arrows; and movement speed by the size of the arrows. Through this screen the *Command & Control Staff* can identify the risk of disturbance in the behaviour of a crowd early on (density too high, pressure against physical barriers, turbulence) and then take the appropriate measures to avoid or mitigate an incident. When the first incident report is received, the Emergency Response Toolkit (ERTK) presents the *Command & Control Staff* with a message indicating the start of an emergency and the specific type of the incident causing the

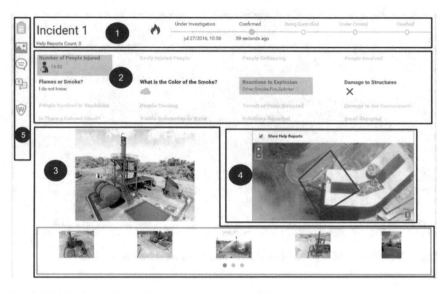

Fig. 5 Visualisation of the incident summary information

emergency. By interacting with this message, the ERTK user has access to the incident summary screen (Fig. 5).

The main components of the screen are delimitated by red frames and identified with numbered red circles. An incident timeline (Circle 1) informs the current status of the incident, indicating the time when the status was achieved. This screen component also allows the ERTK user to update the current status of the incident. In addition, the system automatically changes the incident status to "confirmed" after receiving the first report from a workforce. The aggregated data area (Circle 2) shows the most relevant information obtained from the crowd within the last timeframe. This screen component uses a matrix-based metaphor in which a piece of information is represented in each cell. The reliability of each piece of information is indicated through the colours of the respective information header, where red represents fully reliable information (i.e., confirmed by at least one member of the workforce), orange represents partially reliable information (i.e., coming from at least one member of a supporting force) and green represents information to be confirmed (i.e., coming only from civilians). In order to inform users about information changes, the ERTK highlights the pieces of information that have been changed by painting the background of the respective cells in blue (e.g., Reaction to Explosion in Fig. 5). Changes can be in terms of new values for the pieces of information or in terms of an increase in information reliability. The user can mouse over a highlighted cell to show that they are aware of the change and then the blue background is removed. Furthermore, there is a chart for each piece of information that shows the exact distribution of answers per profile. The ERTK user can access any of these charts by passing the mouse over the respective cell. The incident

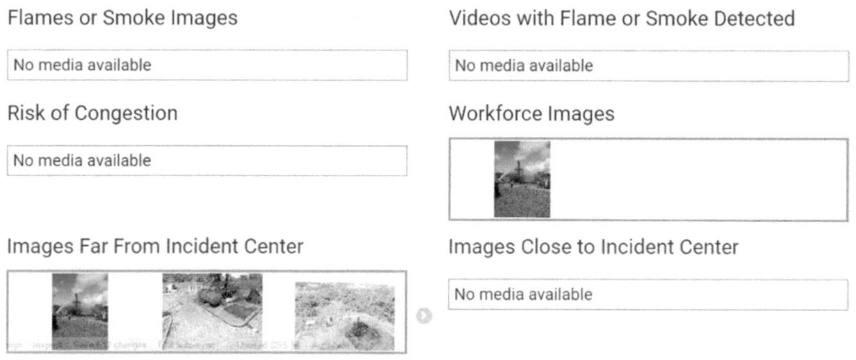

Flames or Smoke Images

> No media available

Videos with Flame or Smoke Detected

> No media available

Risk of Congestion

> No media available

Workforce Images

Images Far From Incident Center

Images Close to Incident Center

> No media available

Fig. 6 Dedicated media view

Fig. 7 Aggregated text information in the form of word clouds

summary screen also presents a media carrousel with images and videos obtained from the crowd (Circle 3). The results of the image and video analyses are used to highlight, hide or order information in the media carrousel. For example: images and videos are ordered according to the presence of anomalies (e.g., fire or crowd disturbance), and image duplicates are hidden. In addition, an incident map (Circle 4) presents critical information related to the incident area. This screen component can show the position of reports sent from civilians that are within the hot zone of the incident (no civilian should stay in this zone), the position of help reports sent from anywhere in the incident area, the position of reports coming from workforces, as well as the places where videos indicate crowd disturbance. Finally, Circle 5 has icons that are used to navigate to other views. The first one (at the top) is the home icon, which a user can use to return to this main view. The second icon opens the dedicated media view (Fig. 6), which presents images and videos sent by the crowd that are specially relevant due to their characteristics (e.g., due to the profile of the sender or their distance from the incident position) or due to anomalies identified by the image/video analyses (e.g., presence of flame or smoke, risk of crowd disturbance). The third icon opens the text analysis view (Fig. 7), which presents the text analysis results for the W^4 questions ("what", "where", "when" and "who") in the form of a word cloud. The fourth icon opens the query and

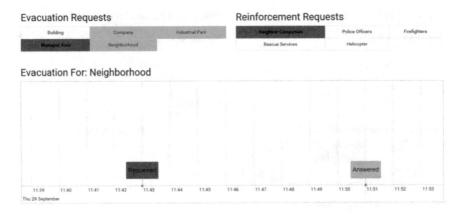

Fig. 8 Query and update reports sent by workforces

update reports view (Fig. 8). This view shows queries of tasks or resources that are still pending (red boxes) and the ones that have already been delivered (green boxes). When a query report is received from a workforce, the boxes corresponding to the requested tasks/resources turn red to get the attention of the *Command & Control Staff*. Red boxes turn green when an update report is received informing the performance of the respective task or the delivery of the respective resource. The timeline in this view presents the moment when the query of a certain task/resource was made and the moment when an update report informed its delivery. Finally, the last icon ("W" letter) is used to show the most recent reports from workforces, so that the ERTK user can easily reach them. If the ERTK user wants to look at a specific report, such as a report from a member of the workforce or from a civilian trapped in the hot zone of the incident, he can access the report through the "W" letter or through the report icon on the incident map. The report view presents all information contained in the report (structured data, video, images, text), and allows the command centre to send requests to the person that sent the report.

4.8 Further Features

In addition, the RESCUER system offers further features to support the command centre in controlling emergencies.

- Public Communication, for the semi-automatic creation of announcements of emergencies to affected communities and the general public. The respective RESCUER component [33] creates announcements from the most appropriate templates taking into consideration the current *Emergency State* and the specific target audience;

- Legacy System Connector, which uses an ontology [34] as common vocabulary for receiving incident reports from sensor-based alarm systems, which is especially relevant in the application scenario of industrial areas, and for directly sending a rich, but concise description of the incident to operational forces' information systems;
- Crowd Guidance, for providing a direct way of informing people in the affected areas about what is going on and how to behave safely. Guidance messages include the minimum necessary information about the incident (what, when and where) and propose clear, small, and simple actions.

5   The Evaluation of the RESCUER Solution

To evaluate the RESCUER system as early as possible and benefit from user feedback in the iterative and incremental development process of the system, we adopted the evaluation strategy illustrated in Fig. 9. This evaluation strategy focused on evaluations involving end users, as the ultimate quality of a software system is indicated by its quality in use. Moreover, the information presented in the Emergency Response Toolkit (ERTK) is the consolidation of the results of all internal system components, which means our strategy allowed end users to indirectly evaluate the results of the internal system components through the evaluation of the ERTK.

We complemented our evaluation strategy with a load test of the overall solution and with individual evaluations of the internal system components. References to the descriptions of these individual evaluations are provided in the respective previous sections. They include data about precision, recall and F-measure.

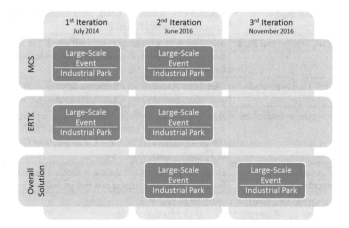

Fig. 9 Evaluation strategy

Concerning the load test, the goal was to evaluate time behaviour, resource utilisation, capacity and availability. A total of 9000 reports were sent to the RESCUER backend. The results showed that the system's response time when navigating through the ERTK screens was not affected by the number of reports, and the average memory and processor consumption were 15 and 1%, respectively. All the time, the message broker used as middleware in the RESCUER system received and sent reports almost in real time. However, after 2000 reports, the ERTK stopped showing reports and the *Emergency State Builder* needed considerable time to consume the reports sent by the load test tool. The partners already have some ideas on how to solve these problems.

Table 1 describes the evaluation exercises performed according to the evaluation strategy in Fig. 9.

As can be observed, there was neither an evaluation of the ERTK in an industrial area nor an evaluation of the overall solution in the second project iteration. In fact, the evaluation exercise identified as "2nd Iteration/MCS/Industrial Park" in Table 1 was supposed to allow not only the evaluation of the MCS, but also the evaluation

Table 1 Evaluation exercises with end-users

Iteration/Scope/Application domain	Description (with number of participants between parentheses)
1st iteration/MCS/Large-scale event [35]	Potential eyewitnesses in the context of the 2014 FIFA Football World Cup: Fritz-Walter-Stadium, Kaiserslautern, Germany (50); FIFA Fan Fest, Salvador, Brazil (35); USP Campus, São Carlos, Brazil (27)
1st iteration/MCS/Industrial Park	Employees of two companies located at the Industrial Park of Camaçari, Brazil (60)
1st iteration/ERTK/Large-scale event & Industrial Park	Command and control centres: CICC[a], Salvador, Brazil (2); CISEM[b], Madrid, Spain (1); Austrian Red Cross and Fire Brigade (5); Camaçari Industrial Park, Brazil (2); Chemical Park of Linz, Austria (1)
2nd iteration/MCS/Large-scale event [36]	Visitors of CIDEM II[c] (incl. workforces and supporting forces), Salvador, Brazil (31)
2nd iteration/MCS/Industrial Park	Simulation at a training centre with real fire (civilians, supporting forces and workforces) at the Industrial Park of Camaçari, Brazil (24)
2nd iteration/ERTK/Large-scale event	Members of emergency response organisations: CIDEM II, Salvador, Brazil (16); CEIC[d], Porto Alegre, Brazil (9)
3rd iteration/Overall solution/Car accident	Real simulation exercise carried out as a pre-condition to the opening of the Lambach tunnel, Lambach, Austria (16 out of 85) with a RESCUER Command and Control Centre at Fireserv, Linz, Austria (4)
3rd iteration/Overall solution/ Large-scale events	Potential eyewitnesses during a football match at the Stadium of Linz, Austria (16 out of 500)

[a]Integrated Command and Control Centre of Bahia, Brazil
[b]Integrated Security and Emergency Centre of Madrid, Spain
[c]Second International Congress on Mass Disasters, held in June 2016, in Salvador, Brazil
[d]Urban Command Centre of Porto Alegre, Brazil

of the ERTK and of the overall solution. However, the RESCUER communication middleware (an industry standard component) became temporarily unavailable; in particular, it refused connections, most probably due to an inappropriate configuration. The delay between the start of the simulation of the fire emergency at COFIC's Training Centre and the solution of the problem made the use of questionnaires for evaluating the ERTK and the overall system not meaningful on this occasion. In addition, there was no evaluation of the overall solution in an industrial area in the third project iteration due to time constraints. This evaluation was replaced by a final system demonstration to potential user organisations at COFIC's Training Centre as part of the final technical project review.

The main results of the performed evaluations were:

- Civilians using the Mobile Crowdsourcing Solution (MCS) gave very positive feedback regarding its usability and would use the application to help the command centre handle emergencies as well as to request help for themselves or for others. Domain experts (members of a workforce or supporting force) also gave very positive feedback about the application and considered it to be suitable for emergencies. However, a considerable number of participants felt they were running risks using the application in at least one of the scenarios that were presented. This led to the creation of a video to instruct potential users on how to use the application taking into account the level of risk posed by different incident scenarios.

- Concerning the ERTK, the focus on mechanisms for the visualisation of aggregated incident data and indication of their reliability was confirmed as being the right approach. This decision was taken in the second project iteration and resulted in very positive evaluation results, with most participants successfully performing all tasks and spending no more than one minute to conclude each task. In addition, valuable feedback was received, such as the wish to assign names and level of importance to incidents and seeing answers to information requests in a more explicit way.

- Regarding the two evaluations of the overall solution, the number of participants was not large taking into consideration the size of the events. However, it is necessary to take into consideration that people were working during the simulation of the car incident and, in the case of the football match, they were there to watch the football match, have fun and talk to friends. Nevertheless, the RESCUER system was demonstrated and evaluated in two real scenarios: In the simulation of a car accident, 276 reports were received by the command centre, which included 2 videos, 25 images and 12 texts. In the football match, 106 reports were received by the command centre, including 9 videos, 49 images, and 6 texts. In both cases, the number of received reports includes updates to a previously sent report. Those numbers are more than sufficient to allow better-informed decision-making in emergencies. In addition, we are confident that the number of people using the MCS will increase considerably when the application is available for download in a mobile application store.

6   Conclusion

This chapter presented RESCUER, a decision support system for emergency and crisis management based on mobile crowdsourcing information. The aim of the system is to make the work of command centres and operational forces in emergencies more effective and efficient by engaging eyewitnesses (i.e., civilians or supporting forces) and formal responders (i.e., workforces) in quickly providing contextual information at the beginning of an incident and keeping the flow of information up until the situation is normalized. In addition to the collection and aggregation of mobile crowdsourcing information, the system has the purpose of supporting official and accurate announcement of emergencies to the affected communities and the general public. When compared to related work, the RESCUER system advances the state of the art because it includes: (1) a mobile application with user interaction mechanisms especially developed for use in emergencies; (2) data analysis capabilities for the automatic detection of fire, smoke and disturbance in crowd behaviour and (3) views on relevant aggregated data such as the incident summary and the view for monitoring the status of queries for tasks and resources.

The RESCUER system was developed in close cooperation with emergency response organisations and potential end-users. A total of 266 civilians, 12 supporting forces, 29 workforces and 30 command and control staff participated in the evaluation exercises. The different evaluation settings offered opportunities for: (1) collecting feedback and evaluating quality attributes in several application scenarios from the viewpoint of different types of stakeholders and (2) acquiring domain knowledge and checking solution approaches in practice in order to continuously develop and improve our system.

However, we have not reached the end of the road. Future development work includes the deployment of an iPhone version of the mobile application and the development of user interfaces to facilitate the configuration of the system for different organisations. The performance issues of the Emergency State Builder also need to be tackled. Regarding future research work, our components for data prioritisation and for video analysis of crowd aspects need a more systematic evaluation. A broader use of open data and domain ontologies should be investigated. Moreover, further image/video analysis components need to be developed to support automatic tagging in the case of other types of incidents. Finally, it is important to investigate whether data analysis components can benefit from the emergency state information captured in the system to optimise their processing.

Acknowledgements The work reported in this paper was carried out in the RESCUER project, a European-Brazilian collaborative project funded by the European Commission (Grant: 614154) and by the Brazilian National Council for Scientific and Technological Development CNPq/MCTI (Grant: 490084/2013-3).

References

1. United Nations Department of Humanitarian Affairs.: Internationally Agreed Glossary of Basic Terms related to Disaster Management. Technical report (1992). http://reliefweb.int/sites/reliefweb.int/files/resources/004DFD3E15B69A67C1256C4C006225C2-dha-glossary-1992.pdf. Accessed 15 July 2017
2. U.S. Department of Homeland Security.: National Incident Management System. Technical report (2008). https://www.fema.gov/pdf/emergency/nims/NIMS_core.pdf. Accessed 15 July 2017
3. BMI (German Federal Ministry of the Interior).: Auskunftsunterlage Krisenmanagement, p. 222 (2011)
4. Engelbrecht, A., Borges, M., Vivacqua, A.: Digital tabletops for situational awareness in emergency situations. In: 15th International Conference on Computer Supported Cooperative Work in Design, pp. 669–676. IEEE (2011)
5. Jolie, K.: Love Parade Duisburg, July 24, Multiperspective-video (2011). https://www.youtube.com/watch?v=up95bUU3L0M. Accessed 14 July 2017
6. Villela, K., Breiner, K., Nass, C., Mendonca, M., Vieira, V.: A Smart and reliable crowdsourcing solution for emergency and crisis management. In: IDIMT 2014. 22nd Interdisciplinary Information Management Talks: Networking Societies—Cooperation and Conflict, Poděbrady, pp. 213–220 (2014)
7. CRISMA—Modelling Crisis Management for Improved Actions and Preparedness (2013). http://www.crismaproject.eu/index.htm. Accessed 14 July 2017
8. Clausthal, T.U.: Rettungsassistenzsystem für Katastropheneinsätze (2011). http://www2.in.tu-clausthal.de/~Rettungsassistenzsystem/. Accessed 14 July 2017
9. Wu, A., Convertino, G., Ganoe, C., et al.: Supporting collaborative sense-making in emergency management through geo-visualization. Int. J. Hum Comput. Stud. **71**(1), 4–23 (2013)
10. Tomoyuki, I., Akira, S., Noriki, U., et al.: A unified large scale disaster information presentation system using ultra GIS based tiled display environment. In: 15th International Conference on Network-Based Information Systems, pp. 550–555. IEEE (2012)
11. Kilgore, R., Godwin, A., Davis, A., et al.: A Precision Information Environment (PIE) for emergency responders: providing collaborative manipulation, role-tailored visualization, and integrated access to heterogeneous data. In: HST'13. 2013 IEEE International Conference on Technologies for Homeland Security, pp. 766–771. IEEE (2013)
12. Poblet, M., García-Cuesta, E., Casanovas, P.: Crowdsourcing tools for disaster management: a review of platforms and methods. In: Casanovas, P., Pagallo, U., Palmirani, M. et al. (eds.) AI Approaches to the Complexity of Legal Systems. Lectures Notes in Computer Science, vol. 8929, pp. 261–274. Springer, Berlin
13. Rogstadius, J., Vukovic, M., Teixeira, C., et al.: CrisisTracker: crowdsourced social media curation for disaster awareness. IBM J. Res. Dev. **57**(5), 4:1–4:13 (2013)
14. Sahana Software Foundation.: Sahana Home of the Free and Open Source Disaster Management System (2012). http://www.sahanafoundation.org/about. Accessed 14 July 2017
15. Heinzelmann, J., Waters, C.: Crowdsourcing Crisis Information in Disaster-Affected Haiti. Special Report, United States Institute of Peace (2010)
16. Zook, M., Graham, M., Shelton, T., et al.: Volunteered geographic information and crowdsourcing disaster relief: a case study of the Haitian earthquake. World Med. Health Policy **2**(2), 7–33 (2010)
17. Ushahidi (2017). https://www.ushahidi.com/. Accessed 14 July 2017
18. Newman, S.: Building Microservices. O'Reilly Media ISBN 10:1-4919-5035-8 (2015)
19. Nass, C., Breiner, B., Villela, K.: Mobile crowdsourcing solution for emergency situations: human reaction model and strategy for interaction design. In: 1st International Workshop on User Interfaces for Crowdsourcing and Human Computation, held at AVI 2014, Como (2014). http://www.st.ewi.tudelft.nl/~bozzon/CrowdUI2014Papers/crowdui2014_submission_5.pdf

20. Luqman, F., Sun, F., Cheng, H., et al.: Prioritizing data in emergency response based on context, message content and role. In: 1st International Conference on Wireless Technologies for Humanitarian Relief, pp. 63–69. ACM (2011)

21. Fajardo, J., Yasumoto, K., Ito, M.: Content-based data prioritization for fast disaster images collection in delay tolerant network. In: 7th International Conference on Mobile Computing and Ubiquitous Networking, pp. 147–152. IEEE (2014)

22. GATE: General architecture for text engineering. http://gate.ac.uk. Accessed 15 July 2017

23. RESCUER Project.: D3.2.3 Data Analysis Method Description 3. Project Deliverable (2017). http://143.107.183.136/?page_id=11037. Accessed 14 July 2017

24. Chino, D., Avalhais, L., Rodrigues, J. Jr, et al.: BoWFire: Detection of Fire in Still Images by Integrating Pixel Color and Texture Analysis. In: SIBGRAPI 2015. 28th Conference on Graphics, Patterns and Images, Salvador, pp. 95–102 (2015)

25. BoWFire Image Dataset.: University of São Paulo, São Carlos Campus (2016). http://gbdi. icmc.usp.br/en/projects/#/projects/2016-bowfire-agma. Accessed 15 July 2017

26. Cazzolato, M., Bedo, M., Costa, A., et al.: Unveiling smoke in social images with the SmokeBlock approach. In: 31st ACM Symposium on Applied Computing, Pisa, pp. 1–6. ACM (2016)

27. Zauner, C.: Implementation and benchmarking of perceptual image hash functions. Master's thesis, Upper Austria University of Applied Sciences (2010)

28. Avalhais, L., Rodrigues, J. Jr, Traina, A.: Fire detection on unconstrained videos using colour-aware spatial modelling and motion flow. In: ICTAI 2016. 28th IEEE International Conference on Tools with Artificial Intelligence, San Jose, pp. 1–8. IEEE (2016)

29. Dalal, N, Triggs, B.: Histograms of oriented gradients for human detection. In: CVPR'05. 2005 IEEE Computer Society Conference on Computer Vision and Pattern Recognition, vol. 1, pp. 886–893. IEEE (2005)

30. INRIA Person Dataset (2006). http://pascal.inrialpes.fr/data/human/INRIAPerson.tar. Accessed 15 July 2017

31. Zach, C., Pock, T., Bischof, H.: A duality based approach for realtime TV-L 1 optical flow. Joint Pattern Recognition Symposium, pp. 214–223. Springer, Berlin (2007)

32. Pérez, J., Meinhardt-Llopis, E., Facciolo, G.: TV-L1 optical flow estimation. Image Process. Line 3, 137–150 (2013)

33. Pereira, J., Novais, R., Vieira, V., et al.: RESCUER news: a public communication tool for crisis situations. In: 1st Workshop on Collaboration and Decision Making in Crisis Situations, held at ACM CSCW 2016, San Francisco (2016)

34. Barros, R., Kislansky, P., Salvador, L., et al.: EDXL-RESCUER ontology: conceptual Model for semantic integration. In: ISCRAM 2015. 12th International Conference on Information Systems for Crisis Response and Management, Kristiansand (2015). http://idl.iscram.org/files/ rebecabarros/2015/1183_RebecaBarros_etal2015.pdf. Accessed 30 Sept 2017

35. Holl, K., Nass, C., Villela, K., Vieira, V.: Towards a lightweight approach for on-site interaction evaluation of safety-critical mobile systems. In: 13th International Conference on Mobile Systems and Pervasive Computing, Quebec. Procedia Computer Science, vol. 94, pp. 41–48. Elsevier (2016)

36. Holl, K., Nass, C., Vieira, V., Villela, K.: Safety-critical mobile systems—the RESCUER interaction evaluation approach. J. Ubiquit. Syst. Pervasive Netw. 9(1), 1–10 (2017)

Intelligent Decision Support for Unconventional Emergencies

Sarika Jain

Abstract This chapter is a conceptual writing speaking about various Decision Support Systems (DSS) available for emergency management and provides a new perspective that is based on the best practices for developing a DSS. There is a requirement of creating citywide situational awareness and its emergency management, which helps its various users, designated as experts in disaster management and city personnel/planners in taking prompt decision in the state of various emergencies. This chapter focuses upon effective and efficient storage, analysis and processing of emergency information, safety plans and resources; and applying an integrated approach of rule base reasoning and case base reasoning to generate the recommendations in case of emergency along with proper justification; hence minimizing the loss of life and property. An ontology representation scheme has been used to represent human knowledge and reason with it. The actions/recommendations are determined based on the historical data (case base) and actions taken, with its real-time synthesis; and validated through the use of existing rules (rule base).

Keywords Decision support · Emergencies · Ontology · Rules
Hybrid reasoning

1 Introduction

Emergency Situations (Disasters) are the situations that occur without warning and can cause widespread distress. An emergency is not always necessarily a disaster but an imminent actual event that threatens people, property, or the environment. They are by nature risky, complex, and unstructured; which has no single correct answer; hence demanding for advisory system or decision support system that

S. Jain (✉)
Department of Computer Applications, National Institute of Technology,
Kurukshetra 136119, Haryana, India
e-mail: jasarika@nitkkr.ac.in

© Springer International Publishing AG 2018   199
R. Valencia-García et al. (eds.), *Exploring Intelligent Decision Support Systems*,
Studies in Computational Intelligence 764,
https://doi.org/10.1007/978-3-319-74002-7_10

provides expertise for making decisions to many problems. Two major types of emergency situations are the natural (earthquake, flood, volcanic eruption, cyclone, tsunamis, epidemics) and the man-made (train accident, building collapse, air crash, bomb blast, warfare) emergencies. The natural disasters (caused by nature) are ad hoc in nature, unpredictable and destructive. Man-made disasters are the events caused due to carelessness of human beings or mishandling of dangerous equipment deliberately or by negligence. Many disasters are not even hard to manage but also very difficult to predict. However, if these disasters are managed in a better way, the accompanying penalties are abridged. We can't stop natural phenomena from happening but can employ cutting-edge information technology (IT) tools to anticipate and minimize the catastrophic impact of natural disasters. Mitigation efforts such as awareness, preparedness, education, prediction systems and recommender systems can prevent or reduce the disruptive impacts of disasters.

The concept of Advisory System or Decision Support System (DSS) picked up emphasis in the middle of 1970s. Decision Support Systems assist human beings in taking decisions by providing advices. They do not take final decisions but provide only expertise to decision makers by automating various tasks of the process. DSS offers potential to assist in solving both unstructured and semi-structured problems with different decision frequencies from low (What actions to be taken) to high (What resources are required). Decision-making often involves the consideration and assessment of situations that do not yet exist and events that have not yet happened. Exploring and scrutinizing such situations requires an abstraction of reality rather than the reality itself. The reality needs to be modeled in narrative, physical, graphical or symbolic way. Artificial intelligence (AI) and Knowledge Management (KM) techniques are extensively employed in all the components of DSS, such as in the knowledgebase, database etc. The inclusion of AI and KM amplifies the thinking capabilities of the expert, makes the tacit knowledge explicit; finds new important facts; integrates both; and detects new patterns and relations along with proper justifications and explanations. Different authors propose different classifications of DSS. No one of them is accepted as a universal taxonomy. In this work we concentrate on knowledge driven DSS which contains stored knowledge and solves problems like human would do. A knowledge driven DSS has two components: a Knowledge Base (Knowledge Modeling and representation technique) and an Inference engine (Reasoning Technique). The advantages of using a knowledge-driven DSS are

- Easy to insert new facts and infer implicit information.
- Able to give explanation and useful for semi-structured and unstructured problems.
- Ontology is used to represent knowledge

  - Highly expressive than database
  - Better when used for large and complex schema.

- Able to represent human knowledge in machine understandable form.
- Better communication and interoperability.

- It provides wide and deep knowledge about a domain, which also includes the explicit rules and semantic association within the area.

Emergency situations are happening frequently and they are a huge danger for human lives and possessions. There is a large variety of emergencies and each of them needs an individual attention. Unconventional emergencies generally lack experiences, and the development of the situation is always dynamic. Unconventional emergency incidents, such as cyclones and earthquakes may and often lead to unforeseen disastrous penalties. The management of such disasters is to be paid highest attention as they directly affect the manpower. In the manual system, on reporting of an emergency, the event is evaluated and a conference of scientific community is called for discussion. A decision on what actions to initiate is taken and communicated to the local public safety officials. Depending on location, kind of emergency and availability of resources the reaction on the situations can be very different. It is comprehensible that planning is essential. Furthermore, a lot of tasks must be coordinated, before the action can take place. Firstly, all necessary information must be gathered and secondly they must be interpreted in the right way. Both of it takes its time but in this case time can decide between life and death of the involved persons.

A bidirectional communication channel between citizens and emergency operators needs to be established so that the humanity is better informed. Through this channel, it would be possible to share updated information about the current situation. On one hand, the citizens can receive alert notifications about the severity and type of emergency, its various attributes, what can be their immediate step, how to make them safe, and any other required information. On the other hand, the operators take advantage of receiving information from the people who are directly involved in the emergency as witness or victim through feedback mechanism. In this way, the citizens can participate actively at the response phase collaborating with the operators for reaching an effective solution as depicted in Fig. 1.

To optimize this process of decision taking we worked on a program, which takes this task over and takes decisions any time and faster than a human being. In this chapter, a recommender system for unconventional emergencies is conceived employing the cutting-edge IT tools and the AI techniques to lower, if not resolve,

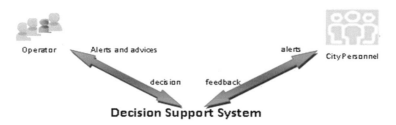

Fig. 1 Bi-directional communication channel

the impacts of the catastrophic disasters. To validate the proposal, focus on one emergency namely earthquake is taken up and all the phases of the decision support are carried out in the context of earthquake.

2 Brief Account of Related Work

In this section, we briefly outline the state of the art literature related to advisory system which provides recommendations to user's query. Specifically, we present relevant literature on traditional emergency response systems and ontology based emergency response systems, particularly in the area of earthquake. A number of international research institutions have done a great deal of work in this emergency area of research. Some researchers have done their work in providing SMS notifications. They include Honeywell,[1] MissionMode,[2] CommandCaller,[3] RapidReach,[4] Sahana,[5] Arce,[6] AlertFind,[7] and Sigame.[8] But they expose weaknesses in giving recommendations to provide decision support in emergency situations especially for earthquake disaster. Although these mentioned systems provide alert service but do not focus on the recipients' specific interests and profiles. The meaningful notifications can be provided by incorporating knowledge base as ontology into the system.

The semantic approach consists in providing affected people with comprehensible and useful information in order to make them aware about what is going on and how to react. In recent years, ontology modeling with semantic tools have been widely used to solve specific modeling and decision support problems in emergency management. Examples of such semantic representations are the Emergency Response Ontology [1] for formalizing the emergency response workflow and the Access Onto ontology [2] for identifying accessible web elements. Reference [3] states the need of an open conceptual and semantic framework to address the emergency response issue. The author emphasizes the development of an open ontology with public consultation. Reference [1] presents a prototype of ontology enabled emergency response system for evacuation. The authors tried to express some generic semantic concepts so that they can be applicable to any crisis. Reference [4] analyzed the incorporation of ontologies in handling semantic interoperability in emergency management.

---

[1]Honeywell: https://instantalert.honeywell.com/.

[2]MissionMode: http://www.missionmode.com.

[3]Command Caller: http://www.voicetech.com/Command_Caller_40.htm.

[4]RapidReach: http://www.rapidreach.com/.

[5]Sahana: http://www.sahana.lk/.

[6]Arce: https://arce.dei.inf.uc3m.es/arce_demo/.

[7]AlertFind: http://www.messageone.com/crisis-communications/.

[8]Sigame: http://www.sigame.es/.

The ENSEMBLE[9] system [5] is a suite of web based software tools, which provides decision support in the management of nuclear emergencies. Recently, National Centre for Seismology launches 'India Quake'—An App for Earthquake Parameter Dissemination. Reference [6] developed the ontology SEMA4A to provide the accessible notifications within emergency response information system (ERIS). Reference [7] presented a system for managing emergency alert notification CAP-ONES which uses SEMA4A (Simple Emergency Alerts 4 [For] All). Reference [8] proposes to notify emergency victims of safe places and available evacuation routes. The authors have extended the SEMA4A ontology and provide customized alerts based on the type of victim, type of emergency and existing infrastructure.

In addition to modeling and representing knowledge in the form of ontology, procedural knowledge support is also required to automate the decision making. Reference [9] proposed a belief rule based (BRB) reasoning approach to assess the suspicion of a disease heart failure by using domain knowledge, signs, symptoms, and risk factors. Reference [10] demonstrated the potential of Web Ontology Language (OWL) reasoning by computerizing paper-based Clinical Practice Guideline (CPG). The authors provided decision support to physicians by employing Semantic web technologies. Reference [11] developed a recommendation system for anti-diabetic drugs selection based on fuzzy reasoning and ontology techniques, where fuzzy rules are used to represent knowledge to infer the usability of the classes of anti-diabetic drugs based on fuzzy reasoning techniques. System/Framework have been built for diagnosis incorporating Case Based Reasoning (CBR) [12], and incorporating hybrid CBR and Rule Based Reasoning (RBR) approach [13]. DSS with RBR exists for improving the company performances and global industrial practices in steel manufacturing [14]. DSHWS [15] is an ontology based system for selecting domestic solar hot water system based on the installation cost, number of occupants, daily consumption etc.

In the domain of disasters, the CBR based fire emergency handling system [16] aims to leverage the information resources to enhance organizational productivity and intelligence. The authors deal with fire emergency by storing previous cases in the form of ontological representation and performing case based reasoning. Reference [17] proposed a novel ontology-oriented decision support system for emergency management based on information integration. The authors tried to work on discovering the effective information and that no piece of information is missed. Reference [18] introduced an Ontology supported intelligent m-Government emergency response services. The authors present a Mobile based Emergency Response System with case-based reasoning approach to infer decision in situations of emergency. The said approach uses ontology for case retrieval algorithm and provides solutions for post disaster events. Reference [19] uses OWL and Semantic Web Rule Language (SWRL) to express the ontologies defining the formal geospatial semantics. They provide a web portal where the users may submit geospatial problems and seek solutions. Reference [20] developed the

---

[9]http://ensemble2.jrc.ec.europa.eu/Web.

meteorological disaster ontology to describe components of the meteorological disaster. The authors proposed a framework, which provides a new method for comprehensive risk assessment of mesh division affected by meteorological disaster. Reference [21] develops a geospatial knowledge repository of disaster information using ontology representation scheme. The idea is that the information remains relevant and the counter measures are operable. Reference [22] presents a unified and formalized plan repository and a prototype of emergency plan training system to facilitate the efficient administrative and operational use of emergency plans. They used an ontology-based knowledge management method and SWRL rules. Reference [23] represented the meteorological disaster system with geo-ontology and demonstrated the analysis results of two interrelated tasks: secondary disaster as an after-effect and evaluation of an evacuation strategy.

3   An Intelligent Decision Support System for Unconventional Emergencies

Many authorities and agencies are working toward disaster management including the Unites States Geological Survey (USGS), the Utah Geological Survey, The International Emergency Management Society (TIEMS), and World Bank to name a few. These agencies have the responsibility of bringing together the national and international humanitarian providers so that they can better contribute towards the collaborated response and reduce the aftermath during emergencies. These agencies are also responsible for optimizing the available resources, conducting research, and coordinating amongst themselves to mitigate the disasters and manage the crisis response. However, there is still a lack of plan and curricula developed to treat and manage unconventional emergencies especially those, which have a repeating nature like earthquakes.

3.1   Research Gaps

Disaster alert apps have become a widely used tool to inform the population about upcoming and ongoing disasters, to inform them about possible consequences, and to instruct them how to behave and which protective measures to take so that the consequences of a disaster can be minimized. On deep analysis of existing literature, it is observed that there is a lack of Knowledge driven advisory system in the context of Earthquake Emergency Situations.

Understanding an earthquake situation; and predicting and acting meticulously are the significant problems that need to be addressed. The different earthquake systems around the globe, the different agencies govt/private/NGOs working in this field, the different information (historical data) available in different non-formalized heterogeneous formats, and the different cultural conventions, all have to be brought on the same platform and integrated. A little or practically no attention has been paid in this direction. Massive store of historical earthquake data is available, but in an unorganized fashion, making it difficult for the decision makers to discover patterns and do any analysis. We propose to use ontology representation scheme to ensure that the semantics are consistent and accessible, and the knowledge gathered is relevant; thereby helping end users in interpreting the notifications correctly; and emergency operators in communicating amongst them.

During reasoning, many researchers have utilized either rule-based reasoning or case based reasoning. Rule base reasoning is monotonic in nature in which truth of a proposition does not change when new information (axioms) is added. Our knowledge about the emergency situation is always incomplete and there must be an exception. We often revise our conclusions, when new information becomes available. There is a lack of non-monotonic reasoning in advisory systems which update the conclusion based on the updated knowledge. Non-monotonic reasoning is able to perform reasoning on incomplete knowledge and handle the exceptions also. There is requirement of a hybrid approach. Many alert generating systems are available in literature but there is requirement of a recommender system which can generate a set of actions at the time of emergency based on previous experiences and the codified expert knowledge.

3.2 Problem Statement

The goal of intelligent decision support for unconventional emergencies is to save life and property of public timely and effectively by assisting decision makers in

- coordinating and commanding emergency response activities,
- providing emergency alerts and recommended actions,
- evaluating risks and selecting appropriate solutions rapidly once emergencies occur,
- providing differentiated services at each response phase to meet requirements,
- restoring critical activities quickly to a safe state thereby reducing the number of damages, victims and casualties.

The proposed system architecture is shown in Fig. 2. The data and processes of the emergency decision support are divided into three vertical layers with the knowledge treasure forming the base of the architecture. Knowledge treasure as defined in [24] serves as a distributed knowledge base which is application independent. Currently we have created our own knowledge treasure which can be linked to the Linked Open Data Cloud (LOD) for maximum visibility and usage. The three layers are input and data pre-processing layer (IDPL), two level computation layer (2LCL) and output and data dissemination layer (ODDL). The sensor data are captured by the input function of IDPL and then passed to the

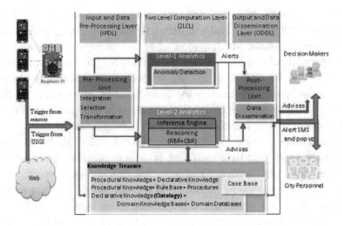

Fig. 2 Schematic of DSS

pre-processing unit for cleaning and formatting. After integration, selection and transformation, the preprocessed data enters the 2LCL which is responsible for anomaly event detection. The 2LCL combines data coming from IDPL with historical knowledge stored in case base and expert knowledge stored in rule base in order to enrich the events semantically; to produce responses to user queries; to facilitate the user with abstract browsing of knowledge base; to generate alerts and advices; and to properly archive the recommendations generated. The ODDL is responsible for disseminating the results to all the registered stakeholders either through email or SMS or pop-ups on their desktop screen. To achieve the said goal, the objectives have been divided into following work elements (tasks):

1. Knowledge Representation and Storage: It involves the development of Declarative Knowledge and the rule base for providing situation awareness.

   - Creating ontology of resources and emergency information, i.e., the domain knowledge base and the domain database. This ontology is named Emergency Situations and Resources Ontology (ESRO).
   - Archiving Past Experience (Storing Cases).
   - Expertise Acquisition (Codifying rules).

2. Generating Recommendations: It is responsible for taking input from the sensors and creating alert messages if an anomaly (unwanted situation) is detected. In parallel, it applies an integrated approach of rule based reasoning and case base reasoning to generate the recommendations along with proper justification. The alerts and recommendations thus generated are also communicated to interested parties with proper access control.

Scientific Importance

1. If not automated, the efficient and effective processing of the earthquake information is not possible.
2. The knowledge treasure will store knowledge of thousands of experts in the rule base and experience of previous historical cases in the case base. Hence all previous experiences and all experts come under one roof.
3. The automated system will be able to generate advises in constrained amount of time.
4. The monotonicity in reasoning of rule based approach has been overcome using the hybrid approach.

3.3 Tools

Various different languages, platforms and developing environments have been utilized for different purposes to achieve the goal. Table 1 summarizes the tools, languages and software used in development phase. All of them are mostly open source in nature.

Programming Language: Java Programming Language has been used for developing all the procedures, say it be parsing the Earthquake information into usable format, case based reasoning, rule based reasoning, or disseminating the results to the stakeholders.

Integrated Development Environment: Netbeans integrated development environment platform is employed and all the user-interfaces are built within.

Data Store: All the historical data of earthquakes, i.e., the cases of the case base are stored in an object-relational data store. PostgreSQL has been employed for the purpose.

Table 1 Tools

SN	Purpose	Languages
1	Programming language	JDK 1.8.0_101, JRE 1.8.0_101
2	Integrated development environment	NetBeans 8.2
3	Data store for storing cases	PostgreSQL 9.6.3
4	Knowledge store for storing resources	Ontology
5	Ontology representation language	OWL2
6	Ontology editor	Protégé 5.2
7	Query language	SPARQL, SQWRL
8	Rule language	Jena, SWRL
9	Application programming interface	API Jena 5.0
10	Semantic reasoning	Java Expert System Shell (JESS)
11	Reasoner	Pellet

Knowledge Store: The earthquake resources are stored in an ontology. An ontology is a formal representation of a set of concepts and the relationships between those concepts.

Ontology Representation Language: Various Ontology languages to represent the knowledge are: SHOE, KIF, RDF(S), DAML+OIL, OWL and OWL2. Required features for a good representation language are: expressive, constraint checking, inference engine, implementation. OWL2 is more expressive and concise to represent the knowledge model. Based on a systematic literature review, we have found that OWL2 has additional features that were missing in OWL and it is an advanced version. OWL2 has been chosen to create the ontology of resources and the earthquake information.

Ontology Editor: Protégé 5.2 is an easy and simple editor to build ontologies created by Stanford University and extensively used by both industrial and academic community for knowledge representation and semantic web applications.

Query Language: SPARQL and SQWRL query language have been used to infer implicit information on ontology with rules.

Rule Language: We provide data inference over SWRL (Semantic Web Rule Language) rules.

Application Programming Interface: Jena API 5.0 has been utilized as it is one of the most popular APIs for creating, storing, managing, querying and manipulating ontologies in Java. Furthermore, Jena provides a query engine in order to execute SPARQL queries over OWL Ontology.

Semantic Reasoning: The Java Expert System Shell (JESS) has been utilized as the rule engine. Jess reasons using the knowledge provided in declarative rules format.

Reasoner: During its evolution, the consistency of ontology is to be maintained. The Pellet reasoner has been used for inferring facts and check consistency.

4 Knowledge Representation and Storage

There is a lack of infrastructure and support for precise, certain, specific and accurate access and retrieval of earthquake information. If the earthquake data from all the available sources are not properly integrated, the agencies may fail in predicting emergency situation/generating of alerts/generating of recommended actions. Databases store flat, large, long-lived data; have multiple users and should satisfy the transactional properties on data. A relational database works upon the closed world assumption and fails if provided with fragmentary or incomplete knowledge [25]. What if the requirement of real world is exactly opposite! What if we need to have data with pointers, no multiple users, the complexity involved in providing transactional properties of data can be avoided; to say, we are able to store, analyze and reuse knowledge. The traditional relational database stores concepts in the form of tables, without containing any information about the meaning of stored concepts or their relationships. Lest the meaning encompassing the concepts and their relationships are stored and understood, any system representing the earthquake knowledge and procedures is incomplete. All types of prediction/notification/alert/recommender systems, may they be available with different terminologies and are made up for different disaster types; must interoperate and communicate with each other in order to share information and provide quick response. The *Knowledge Base* has to be *consistent,* complete, understandable, and interoperable for representing adequately the earthquake knowledge.

The knowledge behind such information has to be modeled as intelligent systems that allow sharing and reusing mechanisms among different IT tools. The role of Information Technology is to support the conceptualization of a knowledge area (i.e. *domain of interest*), making the knowledge formal and explicit for the users. In the literature, several contributions have partly responded to the requirements for designing the *Knowledge Base* component. Different knowledge representation schemes have been proposed in order to cover different organizational needs. To choose a particular representation is crucial to understand what characteristics have to be emphasized or ignored within a topic and to identify which KR technique could capture them. Defined KR techniques can be categorized into three main forms [26]: *Semantic Networks, Production Rules* and *Logic.*

The semantic network is a graph with vertices and edges, where the vertices represent the concepts extracted from the knowledge and the edges represent the relations among the concepts. Each triple over the graph is composed by two vertices and an edge gives a representation of a relevant sentence for the domain of interest.

The production rule is a set of IF-THEN rules for structuring complex statements about a specific knowledge. The advantage of this approach lies in the possibility of deriving easily implicit information about the domain of interest from already defined rules. Censored Production Rule [27], Hierarchical Censored Production Rule [28] and Extended Hierarchical Censored Production Rule [24] came as successors of a production rule by augmenting them with certain operators. An EHCPR is a unit of knowledge resulting in a knowledge base which is modular and hierarchical in nature [29].

The logic consists of representing the knowledge through a formal language that allows creating true or false statements.

Apart from these three categories of KR, recently the notion of ontologies has been adopted as a possible representation for complex domains. Ontologies are powerful semantic models for describing and mapping information related to different knowledge areas. Ontologies provide the means to store such information, which allows for a much richer way to store information [30]. This also means that one can construct fairly advanced and intelligent queries. We use ontology when

(a) we want to represent reality,
(b) we want to simulate human intelligence,
(c) we need to deal with in-complete ever-changing information and large schema,
(d) we want machines to understand the meaning of concepts stored in the knowledge base, we want machines to think like humans and derive new facts,

(e) we want the reasoner of the knowledge base to have more advanced deductive
    capability,
(f) the application requires concept recognition and recommendations,
(g) the source of information is unstructured and heterogeneous,
(h) we want search to be based on open world assumption.

As the knowledge representation scheme, we use an ontology to organize and
store valuable earthquake data. The knowledge treasure thus developed serves as
the store of knowledge and is connected to all the modules of the system.

Knowledge Treasure = *Procedural Knowledge + Declarative Knowledge*

Procedural Knowledge = *Procedures + Rule Base*

Declarative   Knowledge = *Ontology   (Doman   knowledgebase + Domain
database) + Case Base*

The hierarchical network of concepts (domain knowledge base) and the
instances (domain database) are regarded as the ontology of the system. The
earthquake data is very large. For scalable and efficient reasoning it is advisable to
store the archival cases in a relational database. We have chosen PostgreSQL for the
purpose. The rule base stores the knowledge of experts. The rule base and procedures for generating recommendations form part of the procedural knowledge.

In this section, we discuss developing the declarative knowledge (ontology, case
base) and the rule base component of the procedural knowledge:

1. Creating ontology of emergency situations and resources (ESRO), i.e., the
   domain knowledge base and the domain database
2. Archiving Past Experience (Storing Cases)
3. Expertise Acquisition (Codifying rules)

4.1 Creating the Ontology of Emergency Situations and Resources

The process of ontology construction depends on its use. Nevertheless it is an
iterative process. Many methodologies exist in literature for ontology development
e.g., [31, 32] and [33]. We follow a five-phase process to create the ontology of
resources and earthquake information. The five phases are *Scope determination,
Concept Extraction, Concept Organization, Encoding, and Evaluation*. The goal of
scope determination is to list the objectives for which the domain ontology is being
created. Why do we require it? In second phase, we list the core concepts, their
attributes and their instances relevant to the domain. The sources of concepts
extraction should be heterogeneous to get diversified views about the domain. We
have collected concepts from books, internet, by conducting interview, Google
search, brainstorming sessions, journals and conference proceedings. In the third
phase, we arrange the concepts in a hierarchy of generalization and specialization.
Apart from taxonomic (hierarchical) relations, non-taxonomic relations are also

discovered in this phase. In next phase, we encode the hierarchy structure of the previous phase into some chosen representation language and using some ontology development environment. In the last phase, the developed ontology is evaluated for consistency and completeness.

1. **Scope Determination**: The main purpose of developing ESRO is to present a better understanding of the various emergencies. We have prepared a list of competency questions to decide the scope and purpose of ESRO. These questions help us in verifying that the developed ontology contains enough information and that which concepts gathered in phase 2 should be included in the ontology. Some questions listed in this phase include:

   - Display the list of necessary items in First Aid Kit?
   - Which instruments are used to measure the intensity of Natural Disaster X? {Earthquake}
   - Which relief items are provided to persons affected from natural disaster?
   - Display the actions that were taken during the J&K Earthquake.
   - List the various Emergency situations.
   - List the impact factors due to which natural disaster can take place?

2. **Concept Extraction**: Acquire the core concepts and their attributes that attribute to earthquake system. The humanitarian aid information, including information on the occurrences of disaster situations, victims, shelters, resources, facilities, etc., is usually heterogeneous, rapidly changeable, ambiguous, and large. We try to work upon some real-life scenarios and the developed competency questions to come up with a list of concepts to be stored in the ontology.

   The *Event* class includes different kinds of unconventional emergencies along with their damage levels like Rock fall, Drought, Storm, Chemical leakage, Fire, Airspace Accident, Geophysical Disaster, Flood, Tsunami, Heavy Rain, Viral Infections, Avalanche, Heavy Snow, and Earthquake. Every event is associated with certain attributes like: date and time, region, geographical condition, weather condition, rainfall, wind level, traffic condition, flood frequency, disaster reason, disasters level, death toll, emergency level etc.

   The *Action* class includes different kinds of decisions/actions/tasks when the unconventional emergency happens, such as

   - Which roads need to be cleared in order to deploy assistance.
   - Information sharing to ensure a working communication system.
   - Support required for military post and preventive measures.
   - Estimations of required ambulances, required doctors, required temporary hospitals, required medical supplies, required bulldozers/cranes.
   - Positioning/Allocating emergency equipment and supplies.
   - Evacuating potential victims: the location of safe places and available procedures for reaching them, notifying affected people about which routes can be used for escaping.
   - Cleaning of Flooded houses and crops; Snow removing works.

Fig. 3 Concept organization

- Transportation of relief goods like water, medical facility, food, shelter, clothing, cash, communication team, and restoring critical public services. There are various other classes which include the associated data with the unconventional emergency, such as Search and Rescue, Humanitarian Assistance, Location along with the distribution of people, Rehabilitation, and Materials information etc.

3. **Concept Organization**: Organize the concepts in a hierarchy by building relationships among entities (Fig. 3). Concepts extracted in the last phase for Emergency Situations and Resources Ontology are bifurcated into their appropriate categories and converted into a hierarchical structure to represent taxonomic relations. Concepts alone will not provide enough information in ontology. An ontology should contain concepts, their attributes, their instances and exceptions, if any [25]. Properties are used to create the non-taxonomic relationships between concepts (Fig. 4). For example, Earthquake is measured by Seismograph. We have used "measured by" as a property to relate the concepts Earthquake and Seismograph. Top down approach is used to design a hierarchy which represents the concepts from generic term to more specific term.

4. **Encoding**: In this step, we transcribe the hierarchy of concepts into Web Ontology Language (OWL2) documents representing the ontology. Various ontology development tools are available such as Protégé, Apollo and many more. We choose Protégé 5.2 to implement the ESRO. Classes represent the concepts and properties represent the relationship between concepts. There are two types of properties—Object properties and Data properties. Object properties define relationship between classes. Data properties define relationship between classes and values.

5. **Evaluation**: After we are done with the creation of ontology, it is required to be verified and validated. The evaluation and assessment of the ontology is necessary before implementing the application that relies on it. Verification involves checking whether the ontology has been built correctly and validation

Fig. 4 Properties

involves checking whether the correct ontology has been built. ESRO will be called good if it is understood, manipulated and exploited by its intended users. Furthermore, it should be consistent, complete and interoperable for representing earthquake knowledge adequately.

(a) For the *consistency*, we have utilized the Pellet reasoner. Obtained inconsistencies need to be debugged and repaired to get a useful ontology.

(b) Web Ontology Language (OWL) which is a www standard has been used to code ESRO hence guarantying its interoperability. ESRO can be combined with other already existing ontologies and can be easily reused for other purposes.

(c) Completeness and understandability assures that all important information is included in the ontology and the terms and definitions are well named. We have involved various experts in emergency to test the completeness of ESRO. Several domain experts were interviewed and their opinion about the quality of the ontology and the improvements needed were noted and incorporated.

(d) For validation, competency questions are the most effective and reliable way to check if all the important information is included. We have run some complex SPARQL queries for the competency questions and evaluated them for correct answers.

4.2 Archiving Past Experiences (Storing Cases)

Cases can be physically stored in flat files, databases or XML. The DSS creates and stores a large amount of cases, with every case having large number of attributes.

For transaction management, security, integrity and more precisely for simplicity of usage, we have chosen PostgreSQL, an object relational database management system. PostgreSQL is an open source dbms, so it can be used, modified and distributed by anyone for any purpose. Every case (past earthquake event) with its event type, all the required attributes, the calculated emergency level, the decision results/actions proposed by the system, and the actions taken by the authorities are stored in this data store.

4.3 Expertise Acquisition (Codifying Rules)

Knowledge is not always readily available. There are situations where human expertise is scarce, expensive and unavailable (may be in remote locations). Expertise acquisition therefore becomes a cornerstone for DSS as decision quality relies mainly on the quality and availability of acquired knowledge. The ontology layer of the semantic web stack has reached to a reasonable maturity and the focus shifts to the rules layer. Ontology languages don't offer the expressiveness we want. The ontology language OWL is not able to model complex relationships. Whenever there is a requirement of complex semantic involving several concepts, owl needs to be augmented with rules. In the proposed system, first order rules are used to represent the expert knowledge required to take a decision on the emergency. The rules have been added to the ontology using Jena API. Jena can not only read ontologies, but it can be used to reason on ontology the same way that FaCT++ or Pellet can. This can be accomplished using the Jena generic reasoner.

All rules follow the format-

Description or Name of Rule:

$$(\text{condition}_1, \text{condition}_2, \ldots \text{condition}_m) \rightarrow (\text{fact}_1, \text{fact}_2, \ldots \text{fact}_n)$$

An example rule—

If the earthquake has intensity greater than defined critical intensity in particular location, Then "reaction is needed" may be advised:

(?x rdf:type es:GroundShaking), (?x es:hasIntensity ?y), (?x es:hasLocation?c), (?c es:hasCriticalIntensity ?I), greaterThan(?y, ?I) => (?x es:reactionIsNeeded "true").

5 Generating Recommendations

The heart of the proposed system is the recommendation engine, which comprises of three layers vertically with the knowledge treasure lying as bottom-line support to all three of them. The three layers are the input and data pre-processing layer

(IDPL), the two-level computation layer (2LCL) and the output and data dissemination layer (ODDL).

5.1 Input and Data Pre-processing Layer (IDPL)

In this layer, real time input is uploaded by the input resources which are processed and parsed into java object to retrieve the properties of the earthquake. The input resources include the sensor groups installed in the cities and the Email notification service provided by https://www.usgs.gov/. The sensor groups comprise Raspberry pi units with accelerometers which measure the intensity of shaking of the ground and determines the approximate magnitude of the earthquake. The sensor groups are connected to the IDPL to upload the generated report in JSON (JavaScript Object Notation) format. JSON format is a language independent lightweight data interchange format. This report is then processed and parsed at IDPL to obtain the desired properties of the earthquake.

5.2 Two Level Computation Layer (2LCL)

This layer works in two levels and is responsible for generating alerts/advices as the case may be. At level 1, as soon as a trigger is received from IDPL, it is detected whether an anomaly is there and if some action is required. If an anomaly is detected, alert messages are sent to ODDL for further processing and level 2 starts working for generating recommendations. It involves applying an integrated approach of case base reasoning and rule base reasoning to generate the recommendations. Rule Based Reasoning (RBR) and Case Based Reasoning (CBR) can be merged in various forms. The reasoning can start with RBR and heuristically resort to CBR when the applicable rules are contradictory or if the problem to be solved is not an exception to this rule; or it can start with CBR as the main reasoning process and RBR can be used as a partial domain model. In current approach, the case based reasoning represent the past experience in similar type of emergency. It takes into account the past earthquake data and analyses the resources that were provided or needed at that time. The parsed information is analyzed to retrieve the most similar case, keeping in mind all the necessary details and the geographic location. If the similarity value is greater than a threshold value, the actions of the retrieved cases are adapted to meet the current situation with the help of adaptation rules and passed to the ODDL. However, if the similarity value is found out to be less than the threshold, rule based reasoner module is applied and new actions are generated according to the rules defined in the ontology. The working of 2LCL is depicted in Fig. 5.

Multiple scenarios may be constructed by adjusting the parameters. These parameters are: Choosing the location of earthquake, its intensity, defining the number of people in that location, distance between the affected location and the

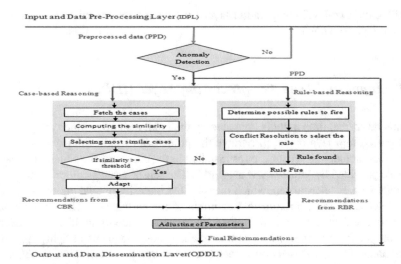

Fig. 5 Working of 2LCL

rescue team. Depending on the rules defined, certain actions will be triggered or other events will be created. The processing of the rule set is done directly on the data in the database. They have the capability to represent the current emergency state and suggest an action or a plan to the operator by the Advisory Subsystem.

5.3   Output and Data Dissemination Layer (ODDL)

This layer is responsible for dispersing the alerts/recommended actions. A message prompt is sent to the Operator on occurrence of an earthquake to login to the site for viewing the recommended actions. The Recommended Actions along with their explanation are displayed online along with the interface with which the operator can make changes to the recommendations and take actions accordingly. The system would also send an Email along with SMS to the required authorities and personnel comprising of details of the Earthquake and few actions that can be taken at that very moment.

6   Concluding Remarks and Future Proposals

Emergency response decisions are complex because they involve multiple organizations, departments, personnel, resources, and legal issues. The fundamental challenges for emergency response decisions are how to make these applications deal with emergencies effectively and how to make victims satisfied during emergency. Artificial Intelligence technologies have come as a rescue.

This chapter is an attempt to integrate the full potential of reasoning approaches as applicable to the domain of unconventional emergencies. State-of-the-art approaches for knowledge representation and storage have been proposed to be used for decision support. The source of information is unstructured and heterogeneous which may lead to complexity of decisions. Such heterogeneity and inconsistency of information sources are easily dealt with by human beings but when it comes to computerized software, it leads to problems. An ontology has been employed as a unified representation of the domain to store the resources and every knowledge about emergencies. Semantics have been embedded into the knowledge and the machine is given the capability to think and infer new facts.

The reasoner module works on the recommendations of history cases and the expert advices. The case based and the rule based reasoning has been beautifully amalgamated to achieve the results. With the aid of such a system, decision makers will be able to take better and timely decisions restoring quickly to a safe situation, hence minimizing the loss of life and property.

Future Proposals

Sky is the limit. Just as the sky seems never ending and infinite, the limit of proposals that technology extends to mankind is as high and wide as the sky. In short run, we perceive to have an Online Web Portal for the system through which an operator can login and review the recommended actions. The Portal should allow the user to feed real-time data in the case base such as number of deaths and injured. Rigorous evaluation and online learning of ESRO is to be done to increase its usefulness. Rules specified in the ontology can be further improved to enhance the precision of the system. Database can be arranged in a way that enables a faster search for finding the similar cases. The system can be generalized to provide advises for other emergencies like tsunami, tornado, etc. Disasters are inherently bound to space and time, but researchers have seldom considered geospatial factors like geographic location specific semantic relationships between concepts.

References

1. Li, X., Liu, G., Ling, A., Zhan, J., An, N., Li, L., Sha, Y.: Building a practical ontology for emergency response systems. In: 2008 International Conference on Computer Science and Software Engineering, pp. 222–225. IEEE (2008)
2. Masuwa-Morgan, K.R., Burrell, P.: Justification of the need for an ontology for accessibility requirements (theoretic framework). In: Interacting with Computers, pp. 523–555. No longer published by Elsevier (2004)
3. Di Maio, P.: An Open Ontology for Open Source Emergency Response System
4. Fan, Z., Zlatanova, S.: Exploring ontologies for semantic interoperability of data in emergency response. Appl. Geomatics. **3**, 109–122 (2011)
5. Mikkelsen, T., Risø National Laboratory.: ENSEMBLE methods to reconcile disparate national long range dispersion forecasts. Risø National Laboratory (2003)

6. Malizia, A., Onorati, T., Diaz, P., Aedo, I., Astorga-Paliza, F.: SEMA4A: an ontology for emergency notification systems accessibility. Expert Syst. Appl. **37**, 3380–3391 (2010)
7. Malizia, A., Acuna, P., Onorati, T., Diaz, P., Aedo, I.: CAP-ONES: an emergency notification system for all. Int. J. Emerg. Manag. **6**, 302 (2009)
8. Onorati, T., Malizia, A., Diaz, P., Aedo, I.: Modeling an ontology on accessible evacuation routes for emergencies. Expert Syst. Appl. **41**, 7124–7134 (2014)
9. Rahaman, S., Hossain, M.S.: A belief rule based clinical decision support system to assess suspicion of heart failure from signs, symptoms and risk factors. In: 2013 International Conference on Informatics, Electronics and Vision (ICIEV), pp. 1–6. IEEE (2013)
10. Jafarpour, B., Abidi, S.R., Abidi, S.S.R.: Exploiting semantic web technologies to develop OWL-based clinical practice guideline execution engines. IEEE J. Biomed. Heal. Inf. **20**, 388–398 (2016)
11. Chen, S.-M., Huang, Y.-H., Chen, R.-C.: A recommendation system for anti-diabetic drugs selection based on fuzzy reasoning and ontology techniques. Int. J. Pattern Recognit. Artif. Intell. **27**, 1359001 (2013)
12. El-Sappagh, S., Elmogy, M., Riad, A.M.: A fuzzy-ontology-oriented case-based reasoning framework for semantic diabetes diagnosis. Artif. Intell. Med. **65**, 179–208 (2015)
13. Zia, S., Akhtar, P., Mala, I., Memom, A.R.: Clinical decision support system: a hybrid reasoning approach (2012)
14. Wang, X., Wong, T.N., Fan, Z.P.: Ontology-based supply chain decision support for steel manufacturers in China. Expert Syst. Appl. **40**, 7519–7533 (2013)
15. Kontopoulos, E., Martinopoulos, G., Lazarou, D., Bassiliades, N.: An ontology-based decision support tool for optimizing domestic solar hot water system selection. J. Clean. Prod. **112**, 4636–4646 (2016)
16. Chakraborty, B., Ghosh, D., Garnaik, S., Debnath, N.: Knowledge management with case-based reasoning applied on fire emergency handling. In: 2010 8th IEEE International Conference on Industrial Informatics, pp. 708–713. IEEE (2010)
17. Han, Y., Xu, W.: An ontology-oriented decision support system for emergency management based on information fusion. In: Proceedings of the 1st ACM SIGSPATIAL International Workshop on the Use of GIS in Emergency Management—EM-GIS'15, pp. 1–8. ACM Press, New York, (2015)
18. Amailef, K., Lu, J.: Ontology-supported case-based reasoning approach for intelligent m-Government emergency response services. Decis. Support Syst. **55**, 79–97 (2013)
19. Jung, C.T., Sun, C.H., Yuan, M.: An ontology-enabled framework for a geospatial problem-solving environment. Comput. Environ. Urban Syst. **38**, 45–57 (2013)
20. Zhang, F., Zhong, S., Yao, S., Wang, C., Huang, Q.: Ontology-based representation of meteorological disaster system and its application in emergency management. Kybernetes **45**, 798–814 (2016)
21. Xu, J., Nyerges, T.L., Nie, G.: Modeling and representation for earthquake emergency response knowledge: perspective for working with geo-ontology. Int. J. Geogr. Inf. Sci. **28**, 185–205 (2014)
22. Luo, H., Peng, X., Zhong, B.: Application of ontology in emergency plan management of metro operation. Proc. Eng. **164**, 158–165 (2016)
23. Zhong, S., Fang, Z., Zhu, M., Huang, Q.: A geo-ontology-based approach to decision-making in emergency management of meteorological disasters. Nat. Hazards **89**, 531–554 (2017)
24. Jain, N.K., Jain, S.: Live multilingual thinking machine. J. Exp. Theor. Artif. Intell. **25**, 575–587 (2013)
25. Jain, S., Gupta, C., Bhardwaj, A.: Research directions under the parasol of ontology based semantic web structure. Adv. Intelli. Sys. Comp. **614**, 644–655, Springer, Cham (2017)
26. Stephan, G. ∈st, Pascal, H. ∈st, Andreas, A. ∈st: Knowledge representation and ontologies. In: Semantic Web Services, pp. 51–105. Springer, Berlin, Heidelberg (2007)
27. Michalski, R.S., Winston, P.H.: Variable precision logic. Artif. Intell. **29**, 121–146 (1986)
28. Bharadwaj, K.K., Jain, N.K.: Hierarchical censored production rules (HCPRs) system. Data Knowl. Eng. **8**, 19–34 (1992)

29. Malik, S., Mishra, S., Jain, N.K., Jain, S.: Devising a super ontology. Proc. Comput. Sci. **70**, 785–792 (2015)
30. Jain, S., Mishra, S.: Knowledge representation with ontology. In: IJCA Proceedings of International Conference on Advances in Computer Engineering and Applications ICACEA. **6**, 1–5 (2014)
31. Corcho, O., Fernández-López, M., Gómez-Pérez, A.: Methodologies, tools and languages for building ontologies. Where is their meeting point? Data Knowl. Eng. **46**, 41–64 (2003)
32. Grüninger, M., Fox, M.S.: Methodology for the design and evaluation of ontologies. In: workshop on Basic Ontological Issues in Knowledge Sharing, Montreal (1995)
33. Pinto, H.S., Martins, J.P.: Ontologies: how can they be built? Knowl. Inf. Syst. **6**, 441–464 (2004)

Information Technology in City Logistics: A Decision Support System for Off-Hour Delivery Programs

Juan Pablo Castrellón-Torres, José Sebastián Talero Chaparro,
Néstor Eliécer Manosalva Barrera, Jairo Humberto Torres Acosta
and Wilson Adarme Jaimes

Abstract City logistics has emerged as one of the major concerns not only for companies but also for policy makers due to its impacts on competitiveness and citizens well-being. Public and private stakeholders have designed strategies such as Off-Hour Delivery (OHD) programs to achieve an equilibrium among economic efficiency, appropriate land use, agile mobility, safety, security and environmental awareness in urban logistics operations. An increasing amount of mega-cities around the world tests, commonly with the government leadership, the benefits of the OHD operations through specific pilots that motivates companies to schedule their deliveries at non-pick hours, e.g. at night. Two of the biggest challenges are making those efforts enduring as well as monitoring OHD operations in the long term. This chapter deploys an information technology strategy based on a decision support system for gathering and sharing data about night deliveries taking as a study case of Bogotá (Colombia). The chapter explains insights of OHD programs, the main indicators designed and the technological architecture that allows the information exchange among public and private actors.

Keywords City logistics · Off-Hour deliveries · Public–private cooperation
Logistics decision support systems · Information technologies · Supply chain
management

1 Introduction

City logistics is a modern concept that describes the optimization process of private companies' transportation and logistics activities in urban areas with the support of advanced information systems. It takes into consideration environmental awareness, congestion, security and energy savings in the framework of economic activities [1].

J. P. Castrellón-Torres (✉) · J. S. Talero Chaparro · N. E. Manosalva Barrera
J. H. Torres Acosta · W. Adarme Jaimes
Universidad Nacional de Colombia, Carrera 30 # 45 – 03
Building 214 of 115, Bogotá D.C., Colombia
e-mail: jpcastrellont@unal.edu.co

© Springer International Publishing AG 2018   221
R. Valencia-García et al. (eds.), *Exploring Intelligent Decision Support Systems*,
Studies in Computational Intelligence 764,
https://doi.org/10.1007/978-3-319-74002-7_11

Besides the companies' process of improving their logistics operations in cities, government from local, regional and national levels appropriated city logistics principles to design public policies that look for people well-being and prosperity. In this wider perspective, in [2] authors define city logistics initiatives as steps taken by municipal administrations to ameliorate the conditions of goods transport in cities and reduce their negative impacts on inhabitants.

Although traffic and transportation activities concentrate the major attention in public decision-making processes [3], those are just a result of supply chain relationships that determine transportation means, frequencies, capacities, schedules, load/unload times, inventory policies, among other logistics decisions exhaustively listed in [4].

In order to successfully face urban logistics challenges from either private or public points of view, solutions have to take into account the whole system with a supply network perspective that includes upstream and downstream actors, third parties, customer preferences, network governance and negotiation schemes [5].

City logistics initiatives that cope with the analysis of supply network dynamics use coordination mechanisms to involve all the actors in win-win situations that make them converge in common goals or objectives including citizens, public authorities and private companies. Arshinder et al. in [6] identify three categories of coordination mechanisms: collaborative initiatives, supply chain contracts and information sharing and technology.

Most of the collaborative initiatives are private companies' strategies to improve their performance in urban logistics activities. Some examples of those are unassisted deliveries [7]; environmental collaboration schemes [8]; joint delivery systems [9]; urban consolidation centers [10]; freight consolidation [11]; cargo bikes for B2B and B2C [12] and delivery points networks for e-commerce [13].

As part of supply chain contracts, there are companies' agreements, in terms of loading/unloading schedules, economies of scale given by buy back, revenue sharing and quantity flexibility schemes [14], that skew practices towards particular efficiency interests under city constraints related to traffic, noise and pollution regulations [15].

Information sharing and technology mechanisms, as Arshinder et al. mentioned, helps to link production nodes all the way until delivery or consumption points. This feature gives a high relevance to this component as an enabler of the implementation of previous coordination mechanisms. Applications on this matter go from private initiatives for sharing logistics information, e.g. Vendor Managed Inventory (VMI) [16], to public platforms for sharing geographical data about traffic regulations, congestion and freight demand/generation zones [17].

Despite initiatives exclusively framed by one mechanism or kind of actor, public and private actors as well as coordination mechanisms converge in programs such as Off-Hour Deliveries (OHD), which make it an interesting project to evaluate in the context of urban logistics.

Along this chapter, we pretend to explain the basis for building an information sharing strategy in OHD programs, based on a Decision Support System that was implemented in Bogotá (Colombia) with the involvement of private companies, associations, academy and government (local and national).

With the aim of providing a guideline to implement the information strategy in the wide specter of OHD programs implementations in megacities worldwide, the chapter begins with an insight of what OHD is. Deployment of information management schemes and DSS architectures are explained in Sects. 3 and 4, respectively. Finally conclusions are drawn.

2 Literature Review

Literature on OHD programs is a recent research field that has emerged with the intent of demonstrating their positive impacts on mobility, sustainability and competitiveness [18]. During the development of this matter, different approaches go in depth with regard to incentives [19], behavioral concerns to induce changes in deliveries schedules [18], economic implementation issues [7], implementation impacts [20] and stakeholder analysis [21, 22] among others. Table 1 shows the scope of some of the most relevant references for OHD programs.

Information technology for gathering data is implicit in all the published papers as a temporal effort made by the research group that headed pilots worldwide. Nonetheless, there is little research on how to design a long-term strategy that involves all the different actors connected in decision support systems for analyzing on-time program performance.

This chapter contributes to the understanding of key performance indicator design and it proposes an architectural scheme for linking actors by a decision support system that monitors the program performance in a long-term period.

3 Insights on OHD Programs—Bogota's Case

Freight transportation impact on cities mobility and vice versa represents a topic of major concern for both policy makers and private companies. People and freight movements converge in the same urban infrastructure that reaches collapsed limits at specific times of the day. Mobility surveys in different countries report catastrophic economic consequences of excessive congestion levels in cities.[1] In Mexico City, data shows that the annual loss caused by being stuck in traffic is about US $1.1 billon [42]. In Buenos Aires, costs are estimated as a 3.4% of GDP [43], and the Inter-American Development Bank IDB calculated an over cost of 120% on

---

[1]Percentage of delay of a trip compared to a free flow situation [46].

Table 1 References in Off-Hour Delivery programs

Description	Reference
Unassisted off-hour deliveries: operations conducted outside regular business hours and without the assistance of the receiving establishment staff	[7, 23]
Economic incentives in OHD programs	[20, 24–26]
OHD implementation cases in cities around the world	[18, 19, 27–31]
Stakeholders analysis	[21, 22, 32–35]
Emissions analysis	[36–39]
Traffic impacts	[20, 40, 41]

average for companies that transport goods in congested areas of cities such as Sao Paulo, Santiago de Chile and Barranquilla [44].

Initiatives to overcome the effects of freight on congestion go from traffic restrictions to volunteering programs that look for separating human movements at peak hours from goods movements inside the city at peak hours [41].

Based on the experiences of different cities around the world, i.e. New York and Barcelona, in 2015 Bogotá realized that the strategy to reduce the amount of trucks in peak hours had to emerge from a joint effort among public, private and academic actors. Regulations were not enough to come up with win-win scenarios where economic (efficiencies for privates), social (mobility improvements) and environmental (emissions reduction) objectives converged in freight operation; therefore, there was a call for a change in a cultural mindset.

A public–private organization leaded the process where local authorities, manufactures and retailers associations made decisions about logistics developments for Bogotá and its surroundings with the support of universities and research groups. The project consisted in making the most of the available infrastructure at night times for freight transportation in order to reduce high congestion levels at day-hours as well as the increasing travel times for private operations.

In the framework of this partnership, Bogotá implemented a pilot with 17 companies from retail, food, petrochemical, beverage and transportation industries that collaborated scheduling their deliveries from 10 p.m. to 4 a.m. in the main industrial areas of the city called Puente Aranda (Industrial Zone 1) and Zona Industrial (Industrial Zone 2) (Fig. 1).

The project main objective was to assess the performance, competitiveness, mobility, security and externalities of a logistics operation, in the implementation of nighttime deliveries during the pilot.

In order to achieve the proposed goal, each company allowed us to collect an amount of information that supported the design of indicators in five categories: environment, safety and security, cost, time and logistics performance. All of them were calculated in daytime as well as nighttime for comparison effects.

Fig. 1 Geographical OHD pilot scope

As pilot general results, average savings on travel expenses registered 32% by comparing a daytime and nighttime distribution operation. Those savings represent an average between heavy trucks (greater than 10 tons) and light trucks (smaller than 10 tons) (Fig. 2).

The cost reduction arose mainly from the traveling time drop, i.e. a route registered 120 min of transit time at pick hours while at night it dropped to 45 min. Additionally, operational times at loading and unloading operations were faster because staff focus on those single operations without interferences generated by customers, providers, administrative duties, etc. In terms of indicators, there was a 20% reduction in operation time of loading in the cargo generator enterprise and 60% savings of unloading times in the cargo receiver enterprise.

Regarding environmental indicators, the pilot showed great results in terms of pollution savings. Comparing daytime versus nighttime, there was a 42% reduction in CO emissions, 8% in CO_2 and 1.4% in PM10 (Fig. 3).[2]

Improvements in air quality impacts of freight transportation activities at nighttime resulted from better driving behaviors (stop and starter frequencies), congestion reduction, amount of vehicles through the route and higher vehicle speeds.

---

[2]Tests conducted with the support of the District Department of Environment, Air, Hearing and Visual Quality Branch of Bogotá.

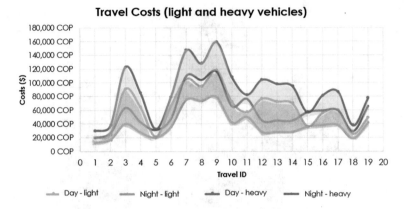

Fig. 2 Travel costs savings for heavy and light trucks

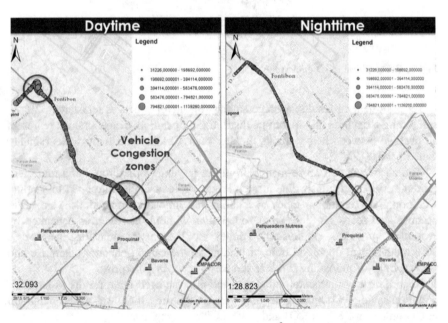

Fig. 3 Nanoparticle concentration in number of particles/cm^3 during a travelling route measured at daytime and at nighttime

Based on the quantitative results and qualitative perceptions from participants (government, private sector and researchers), the list below describes some of the learned lessons from the pilot implementation:

- OHD pilot implementation was cost-effective for the companies involved, where the profit was variable depending on the role in the supply chain.

- Given the regulatory restrictions for transit, loading and unloading in urban centric areas, nighttime operation represents an opportunity to use vehicles with greater capacity.
- More than quantitative results, this first project in Bogotá allowed breaking **the inertia** and misconceptions from private companies about OHD economic and security results.
- The implementation of this project encouraged other companies to participate in upcoming activities.
- Public sector actors can/must work under a coordinated approach.
- These projects must fit into an urban and regional strategic logistics plan that is long term lasting and needs a constant commitment to its implementation.

After the six months pilot, citizens, policy makers and actors from different tiers of supply chains with urban logistics operations in Bogotá did an open call for a long lasting effort for a permanent OHD program.

The challenges identified in Bogotá for attending that call were mainly: (i) to increase the participation and commitment of companies; (ii) to improve safety conditions as well as the perception of the private sector and; (iii) ensuring the supply of transportation services for workers at nighttime.

Nonetheless, designing a robust outreach OHD program requires the identification and synchronization of all players involved in urban deliveries of a specific supply chain i.e. providers, transporters, customers and even consumers [27]. Besides, according to [41], it is necessary to establish the institutional framework by designating human resources focused on freight issues in city agencies or by creating an Industry Advisory Group (IAG).

In the particular case of Bogotá, the foregoing elements cited by authors and practitioners around the globe could exist already. Nonetheless, public–private committees agreed with the need of an information management structure that would enable all the actors to know and analyze OHD program evolution through indicators generated from private companies' data.

The next chapters deploy the information management system fed by indicators and geographical data.

4 OHD Key Performance Indicators

A set of key performance indicators (KPIs), taken from surveys and measurements with technical equipment, such as GPS and sound level meters, were used for each of the companies that participated during the OHD program. Those KPIs belong to five categories, i.e. environment, security, cost, time and logistics performance.

Environmental indicators evaluate emissions, such as noise and pollution, by direct measurements. Security indicators use surveys to analyze perceptions of safety and security in logistics activities. Cost, time and logistics performance indicators are the result of mathematical models described in the following section.

4.1 *KPIs Methodology Calculation*

4.1.1 Logistics Performance KPIs

Queueing Theory is the general framework for calculating KPIs in the context of urban deliveries. The parameters of the model are the following:

λ Number of arrivals per unit time
μ Service rate
NS Number of servers
Wq $E[Tq]$: Average waiting time of customer per queue
Tll Arrival time of trucks
Ti Operation time (hours)
Tto Ending operation time
Tio Starting operation time

Initial relationships for building the model are:

$$Wq(\min) = (Tio - Tll) \times 60 \tag{1}$$

$$\mu(\min) = (Tto - Tio) \times 60 \tag{2}$$

$$\bar{\mu}\left(\frac{pallet}{operator}\right) = \left(\frac{\bar{\mu}}{pallet}\right)/NS \tag{3}$$

$$H = \frac{n}{\sum_{i=1}^{n} \frac{1}{x_i}} \tag{4}$$

where H is the harmonic mean, which is useful for calculating averages with disperse data.

Thus, general logistics performance KPIs are:

a. *Workforce*: it takes into account the average service time per pallet per operator for daytime and nighttime schedules.

$$j = actor; \; k = time\,window\,(daytime\,or\,nighttime)$$

$$H\left(\bar{\mu}(\frac{pallet}{operator})_{jk}\right) = \left[\left[\frac{n}{\sum_{k}^{K} \bar{\mu}\left(\frac{pallet}{operator}\right)_{ijk}}\right]\right] \quad \forall j, k \tag{5}$$

b. *Service time*: These indicators measure the loading and unloading activities time for generators and attractors of freight.

$$(j = actor;\ k = time\ window)$$

$$H\left(\mu(\min)_{jk}\right) = \left[\left[\frac{n}{\sum_{j=1}^{n} \mu_{jk}}\right]\right] \quad \forall k \tag{6}$$

c. *Waiting time*: It allows measuring the general behavior of each actor in terms of lack of efficiency.

$$H\left(Wq(\min)_{jk}\right) = \left[\left[\frac{n}{\sum_{i=1}^{n} Wq_{jk}}\right]\right] \quad \forall k \tag{7}$$

d. *Moved volume*: It refers to the number of pallets moved (EA_{ij}) in loading and unloading operation by each actor in each time window.

$$H\left(VC_{jk}\right) = \left[\left[\frac{n}{\sum_{j=1}^{n} EA_{jk}}\right]\right] \quad \forall k \tag{8}$$

e. *Human resources*: It measures the proportion of the own personnel versus outsourced workforce for loading and unloading activities.

$$NO_{jk} = own\ personnel\ (op)_{jk} + outsourced\ personnel\ (pt)_{jk}$$

$$H\left(NO_{jk}\right) = \left[\left[\frac{n}{\sum_{j=1}^{n} NO_{jk}}\right]\right] \quad \forall k \quad H\left(op_{jk}\right) = \left[\left[\frac{n}{\sum_{j=1}^{n} op_{jk}}\right]\right] \quad \forall k$$

$$H\left(pt_{jk}\right) = \left[\left[\frac{n}{\sum_{j=1}^{n} pt_{jk}}\right]\right] \quad \forall k \tag{9}$$

Table 2 shows the KPI results for the OHD pilot.

4.1.2 Time KPIs

Travel times are the main sources for the time KPIs that depend on speed, route length and time window (daytime or nighttime). Formulations are the following:

$$Route\ lenght = r(\mathrm{km});\ Travel\ time = t(\min);\ Average\ per\ hour = \frac{p}{h};$$

$$Improvement\ percentage = \%m$$

$$\left(\frac{P}{h}\right)_{ij} = \frac{r_{ij} \times 60}{t_{ij}}\ donde\ i = company;\ j = time\ window \tag{10}$$

$$\%m = \left(\frac{r_{i1}}{r_{i2}}\right) - 1$$

Table 2 General logistics performance KPI results in the OHD pilot

General KPI	Actor	Daytime	Nighttime
Average (min)/Pallet	Attractor	9.20	6.90
ū min (pallet/operator)	Attractor	3.07	2.73
Average (min)/pallet	Generator	2.69	2.35
ū min (pallet/operator)	Generator	1.40	0.91
Service time	Attractor	262.08	102.37
	Generator	25.44	20.25
Waiting time	Attractor	55.91	44.21
	Generator	17.10	25.45
Moved volume	Attractor	26.78	14.32
	Generator	12.19	10.96
Human resources	Attractor	3.00	2.40
	Generator	1.53	2.40
Own personnel	Attractor	1	0.5
Outsourced personnel	Attractor	2	2
Own personnel	Generator	1.73	3.00
Outsourced personnel	Generator	0.54	0.50

The ratio %m points out the level of improvement given by higher speeds in the travel time. If %m is greater than one, it means that a nighttime travel produces a better performance in terms of time. Table 3 shows the results for a sample of 10 routes in which nighttime travels have better performance than daytime ones.

General averages for speed and time evaluated in all the participants are:

$$
H\left(\frac{P}{h}\right)_{ij} = \left[\left[\frac{n}{\sum_{i=1}^{n}\left(\frac{P}{h}\right)_{ij}}\right]\right] \quad \forall i,j \quad H(t(\min))_{ij} = \left[\left[\frac{n}{\sum_{i=1}^{n}(t(\min))_{ij}}\right]\right] \quad \forall i,j \quad (11)
$$

Results show that the harmonic mean for daytime speed is 8 km/h, while for nighttime is 16.6 km/h. In terms of travel time, the harmonic average is 84 min in daytime and 46.33 min in nighttime.

4.1.3   Cost KPIs

Cost KPIs design computes resource allocation for loading and unloading activities as well as variable and fixed transportation expenses.

Inputs for these KPI are salaries, transportation cost per kilometer and public services costs per month.

In summary, and after validating workshops with public and private actors, Table 4 shows the group of KPI to monitor in OHD programs.

Table 3 Time KPI results in the OHD pilot

	Time window	Route length (km)	Travel time (min)	Average per hour	%m
Route 1	Daytime	7.6	33	13.82	119.59
	Nighttime	7.5	14.83	30.34	
Route 2	Daytime	25.01	210	7.15	284.21
	Nighttime	21.84	47.73	27.45	
Route 3	Daytime	16.58	157	6.34	268.46
	Nighttime	14.23	36.57	23.35	
Route 4	Daytime	8.9	28	19.07	29.33
	Nighttime	10.52	25.59	24.67	
Route 5	Daytime	10.6	204	3.12	36.32
	Nighttime	9.5	83.9	6.79	
Route 6	Daytime	32.3	221	8.77	28.30
	Nighttime	26.94	130.25	12.41	
Route 7	Daytime	25.75	222	6.96	15.50
	Nighttime	23.73	146.66	9.71	
Route 8	Daytime	32.21	274	7.05	29.48
	Nighttime	28.53	137.21	12.48	
Route 9	Daytime	26.5	59	26.95	1.76
	Nighttime	22.61	49.47	27.42	
Route 10	Daytime	10.23	99	6.20	190.42
	Nighttime	20.56	68.51	18.01	

5 Decision Support System for OHD Programs

After designing the battery of KPIs, the challenge goes toward the design of a Decision Support System (DSS) that allows all the participants to report and have access to data and analysis about OHD program performance. The architecture design has to provide a tool for calculating the KPIs of Table 4.

According to [45], smart cities must implement DSS to support their infrastructure and mobility programs. Among the requirements, the paper establishes building blocks to achieve a successful digital strategy, e.g. ubiquitous connectivity, anytime/anyplace devices, collaboration platforms, cloud computing, open standards and ecosystems, geospatial platforms, Internet of Things, advanced analytics, open access to public data, digitally controlled devices and web 2.0 and social networking.

Besides technological issues, [15] consider that this DSS has to take into account facts as:

- A collaborative process that includes all stakeholders of city logistics and starts from the analysis of the stakeholders needs.
- Dialogue is a key principle that should build trust among all stakeholders.

Table 4 Consolidated list of key performance indicators for OHD

Category	KPI	Actor		
		Generator	Transport	Attractor
Environment	Emissions transport variation		x	
	Noise levels in deliveries		x	x
Security	Number of thefts	x	x	x
Logistics performance	Daytime–nighttime operations proportion	x		x
	Volume moved proportion	x		x
Cost	Workforce cost variation	x	x	x
	Equipment cost variation	x		x
	Public services cost variation	x		x
	Security expenses variation	x		x
	Transportation cost variation		x	
	Congestion variation		x	
Time	Waiting time variation	x	x	x
	Travel time variation		x	
	Operation time variation	x		x

- A process of mutual understanding of the needs and expectations as well as stakeholders involvement must be considered.
- All the stakeholders should be involved in the development of strategic plans, using available tools and methods.

5.1 Solution Architecture

Figure 4 depicts the general DSS structure of the OHD program. The system begins with an odb2 interface in the vehicles that generates information about variables related to speed, emissions and oil–oxygen ratio. An Embedded Tracking Device (ETD)[3] (Fig. 5) with different kind of sensors gathers data of time, temperature, humidity, slope, position and indirect speed. Sensors send the data by means of a connection between the ETD and a small single-board computer Raspberry PI 3 that is in charge of compiling, storing. Data is sent through different protocols and interfaces that allow the link of diverse components to the technological solution.

The ETD sends records to the CARTO platform in json format via a REST-FULL interface by executing the application protocol http of the reference model TCP/IP.

---

[3]The Embedded Tracking Device (ETD) is composed of a device with processing and storing capacity that is in charge of providing physical and logical infrastructure for an orchestrator execution that manages a finite state machine.

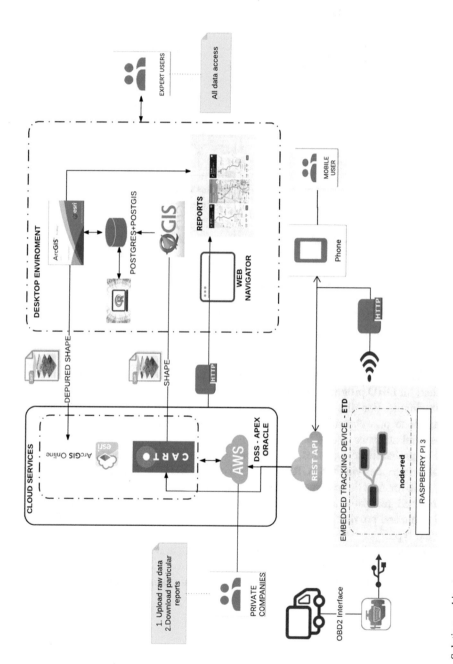

Fig. 4 Solution architecture

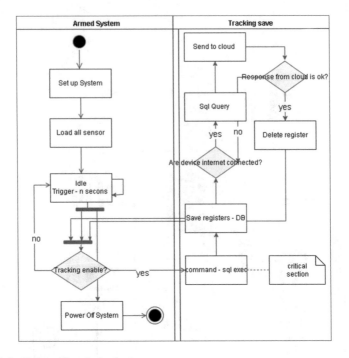

Fig. 5 Embedded tracking device logic

APEX tool from Oracle provides services that support the Decision Support System for OHD programs based on Geographical Information Systems (GIS) as its main input. For the geo-processing analysis execution, those GIS can be stand-alone or web with the aim of obtaining routes calculation, thematic maps, geospatial statistics, etc.

For probabilistic analyses, the system employs R that has access to the geo-database services via POSTGRES and its add-in POSTGIS. Those services are also available for ArcGis desktop that provides the link to ArcGis online. The latter does specialized geospatial analysis such as route calculations based on historical data of real traffic and provides scenarios for best paths according to the time of the day and vehicle type.

For visualization purposes (Fig. 6), an Apex Oracle web interface provides KPI reports,[4] the ArcGis online geoportal generates thematic maps and CARTO allows some visualization options such as animations, heat maps, dynamic filters, among others utilities.

---

[4]A first attempt developed by Imétrica is available in https://app.powerbi.com/view?r= eyJrIjoiMDE0ZTE5ZWQtYmM0ZC00ODFiLTg1Y2ItN2I2ZmM2ZDJm%20OGVhIiwidCI6IjEz MmY0NTdlLTBiMDUtNDNjMi04MWQ1LThiOWY1NGM1ZDMzZiIsImMiOjR9.

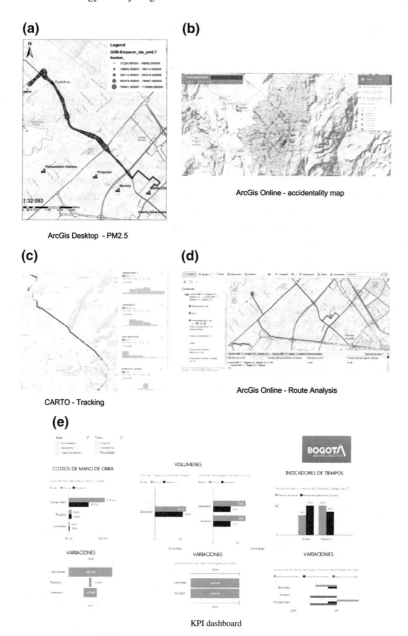

Fig. 6   Visualization print screens of the OHD program web interface

6 Conclusions and Future Research

The chapter deployed an information technology strategy based on a Decision Support System for gathering and sharing data about night deliveries, with the experience of Bogotá (Colombia). The chapter explained the insights of OHD programs, the main designed indicators and the technological architecture that allows the information exchange among public and private actors.

The OHD pilot implementation showed that designing a robust outreach OHD program requires the identification and synchronization of all players involved in urban deliveries and establishing the institutional framework for freight issues in city agencies. Besides, it must satisfy the need of an information management structure that enable all the actors to know and analyze OHD program evolution through indicators generated from private companies data.

A battery of 14 KPI formed the quantitative approach to monitor OHD programs with regard to environment, cost, time, logistics performance and security. Each of those indicators assess data for each participant company and analyze variations between nighttime and daytime operations.

The developed tool allows supply chain actors and policy makers to have access to KPI reports. It also generates thematic maps and some visualization options such as animations, heat maps, and dynamic filters, among other utilities.

Future research fields emerge from this chapter based on the information usage from the DSS. For example, further analyses about program performance via simulation, scenario planning using, and route optimization routines, among others.

References

1. Taniguchi, E., Thompson, R., Yamada, T.: Visions for City Logistics. Presented at the (2004)
2. Awasthi, A., Chauhan, S.S.: A hybrid approach integrating affinity diagram, AHP and fuzzy TOPSIS for sustainable city logistics planning. Appl. Math. Model. **36**, 573–584 (2012)
3. Allen, J., Ambrosini, C., Browne, M., Patier, D., Routhier, J.-L., Woodburn, A.G.: Data collection for understanding urban goods movement: comparison of collection methods and approaches in European countries (2013)
4. Christopher, M.: Logistics and Supply Chain Management: Strategies for Reducing Cost and Improving Service. Financial Times/Pitman, London (1998)
5. Morana, J.: Sustainable supply chain management in urban logistics. In: Sustainable Urban Logistics: Concepts, Methods and Information Systems, pp. 21–35. Springer, Berlin (2014)
6. Arshinder, K., Kanda, A., Deshmukh, S.G.: A review on supply chain coordination: coordination mechanisms, managing uncertainty and research directions. In: Supply Chain Coordination Under Uncertainty, pp. 39–82. Springer, Berlin (2011)
7. Holgu'n-Veras, J., Marquis, R., Brom, M.: Economic impacts of staffed and unassisted off-hour deliveries in New York City. Proc. Soc. Behav. Sci. **39**, 34–46 (2012)
8. Seroka-Stolka, O.: Green initiatives in environmental management of logistics companies. Transp. Res. Proc. **16**, 483–489 (2016)
9. Taniguchi, E.: Concepts of city logistics for sustainable and liveable cities. Proc. Soc. Behav. Sci. **151**, 310–317 (2014)

10. de Correia, V. A., de Oliveira, L.K., Guerra, A.L.: Economical and environmental analysis of an Urban Consolidation Center for Belo Horizonte City (Brazil). Proc. Soc. Behav. Sci. **39**, 770–782 (2012)

11. Castrellón-Torres, J.P., García-Alcaraz, J.L., Adarme-Jaimes, W.: Freight consolidation as a coordination mechanism in perishable supply chains: a simulation study. DYNA **82**, 233–242 (2015)

12. Muñuzuri, J., Cortés, P., Onieva, L., Guadix, J.: Modelling peak-hour urban freight movements with limited data availability. Comput. Ind. Eng. **59**, 34–44 (2010)

13. Cárdenas, I., Sánchez-Díaz, I., Dewulf, W.: Coordination in delivery points networks for the e-commerce last-mile. In: Taniguchi, E. (ed.) The 10th International Conference on City Logistics, pp. 148–161. Institute for City Logistics, Phuket, Thailand (2017)

14. Chanut, O., Paché, G.: Supply networks in urban logistics–which strategies for 3PL? In: International Conference on Economics and Management of Networks, pp. 1–19. Limmasol, Cyprus (2011)

15. Witkowski, J., Kiba-Janiak, M.: The role of local governments in the development of city logistics. Proc. Soc. Behav. Sci. **125**, 373–385 (2014)

16. Disney, S.M., Potter, A.T., Gardner, B.M.: The impact of vendor managed inventory on transport operations. Transp. Res. Part E Logistics Transp. Rev. **39**, 363–380 (2003)

17. Guerlain, C., Cortina, S., Renault, S.: Towards a collaborative geographical information system to support collective decision making for urban logistics initiative. Transp. Res. Proc. **12**, 634–643 (2016)

18. Marcucci, E., Gatta, V.: Investigating the potential for off-hour deliveries in the city of Rome: retailers' perceptions and stated reactions. Transp. Res. Part A Policy Pract. **102**, 142–156 (2017)

19. Brom, M., Holguín-Veras, J., Hodge, S.: Off-hour deliveries in Manhattan, New York City. Transp. Res. Rec. J. Transp. Res. Board **2238**, 77–85 (2011)

20. Holguín-Veras, J., Ozbay, K., Kornhauser, A., Brom, M., Iyer, S., Yushimito, W., Ukkusuri, S., Allen, B., Silas, M.: Overall impacts of off-hour delivery programs in New York City Metropolitan Area. Transp. Res. Rec. J. Transp. Res. Board **2238**, 68–76 (2011)

21. Stathopoulos, A., Valeri, E., Marcucci, E.: Stakeholder reactions to urban freight policy innovation. J. Transp. Geogr. **22**, 34–45 (2012)

22. Ballantine, E., Lindholm, M., Whiteing, A.: A comparative study of urban freight transport planning: addressing stakeholder needs. J. Transp. Geogr. **32**, 93–101 (2013)

23. Holguín-Veras, J., Wang, X. (Cara), Sánchez-Díaz, I., Campbell, S., Hodge, S., Jaller, M., Wojtowicz, J.: Fostering unassisted off-hour deliveries: the role of incentives. Transp. Res. Part A Policy Pract. **102**, 172–187 (2017)

24. Silas, M., Holguín-Veras, J., Jara-Díaz, S.: Optimal distribution of financial incentives to foster off-hour deliveries in urban areas. Transp. Res. Part A Policy Pract. **46**, 1205–1215 (2012)

25. Holguín-Veras, J.: Necessary conditions for off-hour deliveries and the effectiveness of urban freight road pricing and alternative financial policies in competitive markets. Transp. Res. Part A Policy Pract. **42**, 392–413 (2008)

26. Holguín-Veras, J., Pérez, N., Cruz, B., Polimeni, J.: Effectiveness of financial incentives for off-peak deliveries to restaurants in Manhattan, New York. Transp. Res. Rec. J. Transp. Res. Board **1966**, 51–59 (2006)

27. Bertazzo, T., Hino, C., Lobão, T., Tacla, D., Yoshizaki, H.: Business case for night deliveries in the city of São Paulo during the 2014 World Cup. Transp. Res. Proc. **12**, 533–543 (2016)

28. McPhee, J., Paunonen, A., Ramji, T., Bookbinder, J.H.: Implementing off-peak deliveries in the Greater Toronto Area: costs, benefits, challenges. Transp. J. **54**, 473 (2015)

29. Sánchez-Díaz, I., Georén, P., Brolinson, M.: Shifting urban freight deliveries to the off-peak hours: a review of theory and practice. Transp. Rev. **37**, 521–543 (2017)

30. Bentolila, D.J., Ziedenveber, R.K., Hayuth, Y., Notteboom, T.: Off-peak truck deliveries at container terminals: the "Good Night" program in Israel. Marit. Bus. Rev. **1**, 2–20 (2016)

31. Fu, J., Jenelius, E.: Transport Efficiency of Off-Peak Urban Goods Deliveries: A Stockholm Pilot Study (2017)
32. Holguin-Veras, J., Polimeni, J., Cruz, B., Xu, N., List, G., Nordstrom, J., Haddock, J.: Off-peak freight deliveries: challenges and stakeholders perceptions. Transp. Res. Rec. 42–48 (2006)
33. Verlinde, S., Macharis, C.: Who is in favor of off-hour deliveries to Brussels supermarkets? Applying multi actor multi criteria analysis (MAMCA) to measure stakeholder support. Transp. Res. Proc. **12**, 522–532 (2016)
34. Holgín-Veras, J., Campbell, S., Kalahasthi, L., Wang, C.: Role and potential of a trusted vendor certification program to foster adoption of unassisted off-hour deliveries. Transp. Res. Part A Policy Pract. **102**, 157–171 (2017)
35. Jaller, M., Holguín-Veras, J.: Comparative analyses of stated behavioral responses to off-hour delivery policies. Transp. Res. Rec. J. Transp. Res. Board **2379**, 18–28 (2013)
36. Holguín-Veras, J., Encarnación, T., González, C., Winebrake, J., Wang, C., Kyle, S., Herazo-Padilla, N., Kalahasthi, L., Adarme, W., Cantillo, V., Yoshizaki, H., Garrido, R.: Direct impacts of off-hour deliveries on urban freight emissions. Transp. Res. Part D Transp. Environ. (2016)
37. Palmer, A., Piecyk, M.: Time, Cost and CO_2 Effects of Rescheduling Freight Deliveries
38. Wang, X., Zhou, Y., Goevaers, R., Holguin-Veras, J., Wojtowicz, J., Campbell, S., Miguel, J., Webber, R.: Feasibility of installing noise reduction technologies on commercial vehicles to support off-hour deliveries. Submitted to the New York State Energy Research and Development Authority (NYSERDA) and New York State Department of Transportation (NYSDOT) (2013)
39. Campbell, J.F.: Peak period large truck restrictions and a shift to off-peak operations: impact on truck emissions and performance. J. Bus. Logistics **16**, 227 (1995)
40. Iyer, S.: Estimating Traffic Impacts of an Off-Hour Delivery Program Using a Regional Planning Model (2010)
41. Holguín-Veras, J., Wang, C., Browne, M., Hodge, S.D., Wojtowicz, J.: The New York City off-hour delivery project: lessons for city logistics. Proc. Soc. Behav. Sci. **125**, 36–48 (2014)
42. Taiyab, N.: Transportation in Mega-cities: A Local Issue, a Global Question (2008)
43. Ifmo: Megacity Mobility Culture. Springer, Berlin (2013)
44. IDB: Methodology to Analyze and Quantify the Impacts of Congestion on Supply Chains in Latin-American Cities. Washington (2016)
45. Nowicka, K.: Smart city logistics on cloud computing model. Proc. Soc. Behav. Sci. **151**, 266–281 (2014)
46. Kin, B., Verlinde, S., Macharis, C.: Sustainable urban freight transport in megacities in emerging markets. Sustain. Cities Soc. **32**, 31–41 (2017)